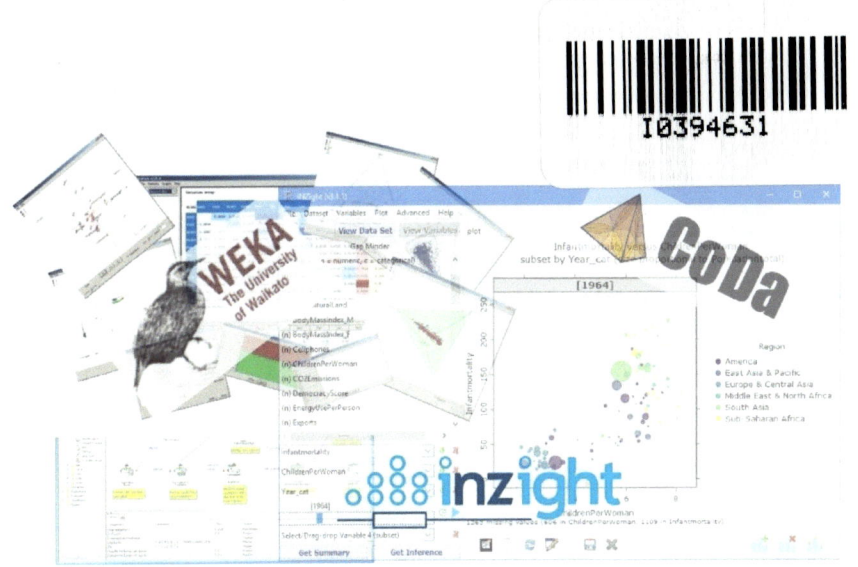

Exploring Geological Data with Weka, CoDaPack, and iNZight

Graphical Instructions

P. GEO. RICARDO A. VALLS

Copyright © 2017 Ricardo A. Valls

All rights reserved. This publication is protected by copyright and permission should be obtained from Valls Geoconsultant prior to any prohibited reproduction, storage in a retrieval system, or transmission in any form or by any means, electronic, mechanical, photocopying, recording, or likewise. For information regarding permission write to Valls Geoconsultant at 1008-299 Glenlake Ave., Toronto, Ontario, Canada, E-mail: vallsvg@gmail.com.

ISBN: 978-1548620332
ISBN-10: 1548620335

DEDICATION

To those who are asking new questions,
searching new ways,
and finding new paths.
To my wife for helping me with the edition and the whole process.
And to my son and family… just because.

Table of Contents

Abstract 2
   Weka 2
   Compositional data analysis 2
   INZight 3
Chapter I   Acquiring the data 4
   Let us get the data first 4
   Stream sediments 4
      When and where to use it 4
      Sampling method 5
      Density of sampling 6
   Heavy mineral concentrates (HMC) 6
      When and where to use it 6
      Sampling method 6
      Density of samples 11
   Soil sampling 11
      When and where to use it 11
      Sampling method 11
      Density of samples 12
   Determining the best sampling method 13
Chapter II   Establishing the quality of the data 15
   Duplicate samples 16
   Blanks 16
   Standard samples 17
   External control 17
Chapter III   Preparing the data 19
   Cleaning the data 19
      When zero doesn't mean it. 19
      Amalgamation 26
      Normalization by Fe and Mn 29
      Estimating the erosional level 30
      Class 31
   What is Weka? 31
      Description 32
      User Interfaces 32
      Extension packages 33

    History 33
   Preparing the Data with Weka 34
    Reducing the size of the dataset...34
Chapter IV Exploring the data...40
   Exploratory data analysis with Weka...40
    Histograms..40
    Removal of statistical outliers 42
    Clustering..42
Chapter V   Using the experimenter in Weka...............................45
Chapter VI Cost estimation with Weka 47
   Evaluating the cost of your prediction 47
   ROC 49
Chapter VII Processing the data 51
   What is CoDaPack? 51
   Compositional data analysis..51
   Biplot to select informative elements..57
   Balance dendrogram 59
   Transforming your original data 62
    ALR 62
    CLR..63
    ILR...64
    "Normal" statistics...67
   Scatterplots...68
   Ternary principal components 69
Chapter VIII Presenting the data 70
   INZight..70
    Mono element representation 71
    Bi element representations..74
    Multielement representations..77
    Time series..80
   Other free programs to consider..81
Chapter IX Conclusions and recommendations 82
References...83
About the author 86

ACKNOWLEDGMENTS

I am indebted to professor Ian Witten of the Waikato University in New Zealand (the top of the World) for the multiple videos and clear presentations of machine learning on YouTube. I am also grateful to professors Vera Pawlowsky Glahn, Juan José Egozcue, and Santi Thió Henestrosa of the University of Gerona for their help and support on my path to understanding better compositional data analysis.

Abstract

Weka

Everybody is talking about Data Mining and Big Data. Data mining turns raw data into useful information. Machine learning is a type of artificial intelligence (AI) that teaches computers without being explicitly programmed. Machine learning focuses on the development of computer programs that can change when exposed to new data. Both Data Mining and Machine Learning search through data to look for patterns.

Weka is a powerful, yet easy to use software, for machine learning and data mining. Weka contains tools for data pre-processing, classification, regression, clustering, association rules, and visualization.

Weka is an open-source software issued under the GNU General Public License. It can be freely downloaded from http://www.cs.waikato.ac.nz/ml/Weka/downloading.html.

Compositional data analysis

The awareness of problems related to the study of compositional data analysis dates back to a paper by Pearson (1896) whose title begin with the words *"On a form of spurious correlation ... "*.

Compositional data (CoDa) are vectors of positive components and a constant sum, e.g. 100% or 1. These conditions render most classical statistical techniques incoherent on compositions, because they were devised for unbounded real vectors. However, many types of data exhibit the same limitations. When the variables of a data set show the relative importance of some parts of a whole, data should be considered compositional.

Examples of disguised compositions are data presented in ppm, ppb, molarities, or any other concentration units. Aitchison (1982) introduced the log-ratio approach for analyzing compositional data. He based his solution on mapping the data vector with some log-ratio transformations and applying classical techniques to the scores thus obtained. This became the foundation of the compositional data analysis (CoDa).

At the present time, CoDa has its own geometric structure for the simplex. The validity of these considerations is unrestricted to CoDa as many more data sets disobey the rules of real numbers, or which can be given an own, alternative, meaningful geometric structure. Examples abound in natural and social sciences, for example vectors of positive amounts, functional data, spherical data, ordered variables, etc.

Practitioners interested in CoDA can find a forum of information, materials, and ideas at www.compositionaldata.com. They can also download from http://ima.udg.edu/codapack the free Compositional Data Package (CoDaPack) that employs the most elementary compositional statistical methods. This software was developed and is maintained by the University of Girona research group on compositional data analysis. It is designed for the applied sciences background users lacking experience using computer packages. Finally, they can also download for free from www.compositionaldata.com the R Packages compositions, robCompositions, and zCompositions (V. Pawlowsky-Glahn & Egozcue, 2006).

INZight

INZight is a free software for statistical data analysis from https://www.stat.auckland.ac.nz/~wild/iNZight/index.php.

While I am discussing here geochemical data, these analytical methods are applicable to other fields of geology, such as geophysics, hydrogeology, etc. Download the data sets to follow these procedures from the web at http://ow.ly/l0Lr30eoSE1.

Keywords: Weka, ML, Coda, CoDapack, Inzight, Machine Learning, Geology, Geochemistry

Chapter I
Acquiring the data

Let us get the data first

You can use geochemical methods in almost all types of environments (regoliths), but the environment and the sought target will define the most appropriate geochemical method to use.

An exploration program can be regional or local. Most of the geochemical methods are cheap so the budget is inconsequential in choosing the method.

Regional programs benefit more from the study of larger mechanical-chemical areoles, so regional methods like stream sediments, heavy mineral concentrates (HMC), biogeochemistry, hydrogeochemistry, atmogeochemistry, and others will be effective in this environment. Local programs should, on the contrary, avoid displaced areolas, so samples from soil and rocks will be favored during more local studies.

The specific method to use in each case depends on the regolith and general morphology of the area. When the area under study lacks a well-developed hydrographic grid, stream sediments and HMC are not recommended.

You may be confronted with one of two situations. Either you have a data set of results, or you need to plan the work necessary to create such dataset.

If you ALReady have a data set please go to "Cleaning the data" section.

Stream sediments

When and where to use it

1. Mostly for regional works.
2. Active streams.
3. Only first or second-order streams.

Sampling method

1. On a topographic map, determine the erosional basins by delineating the water-partition points (wpp). The wpp is the line that joins together the points of maximum elevation within a basin. The basin is call "basin of denudation".

2. Locate the sampling points on the map. I suggest that you take the first three samples in a triangular fashion, with the base of the triangle some 50 metres from the first sample in the creek.

3. Take the samples from first and second-order creeks, at regular spaces if possible looking to characterize the entire basin of denudation. When planning the survey, color code the higher-order streams (first, second, and third). First-order streams are those without tributaries. When a stream meets another stream of the same order, its designation increases by one.

4. When two creeks join, take samples from each creek, but never immediately after their joining.

5. Take samples preferably from below the water level, and always from the same margin of the creek.

6. In the most rudimentary fashion, you can take a sample by putting your hand inside an inside-out plastic bag, grabbing a fistful of the material and pulling the bag to its normal condition. Discard any big rock fragments, roots and other vegetation from the bag.

7. The best sampling procedure is to take the sample with a plastic or stainless steel shovel (Fig. 1) and pass the sample through a -80 Mesh sieve to eliminate all rock fragments and vegetation (Valls, 2013).

Figure 1. Folding shovel from Deakin (http://ow.ly/zvAM30eaMiS).

8. After describing the sample (Valls, 1987), close the bag tightly.
9. If the proper conditions exist, air-dry the samples at camp.
10. If possible, avoid re-bagging.

Density of sampling

For a regional scale of 1:50 000 take 4-6 samples per square km from the first or second-order streams. For country-size projects, you can use densities of 1 sample per square kilometer. Such grid was successfully used during a regional survey of the Amazon river in Brazil. Stream sediments at scales larger than 1:10 000 (approximately 10-12 samples per square km) are inadvisable. At such scale, soil sampling methods with or without a grid are more effective.

Heavy mineral concentrates (HMC)

When and where to use it
1. Mostly for regional works.
2. Active streams.
3. Only first or second-order streams.

Sampling method

The sample selection for HMC sampling is like that of the stream sediments. Both methods look to define areas of mineralization within the basins of denudation.

1. On a topographic map, determine the erosional basins by delineating the water-partition points (wpp). The wpp is the line that joins together the points of maximum elevation within a basin. The basin is call "basin of denudation".
2. Locate the sampling points on the map. I suggest that you take the first three samples in a triangular fashion, with the base of the triangle some 50 metres from the first sample in the creek.
3. Take the samples from first and second-order creeks, at

regular spaces if possible looking to characterize the entire basin of denudation. When planning the survey, color code the higher-order streams (first, second, and third). First-order streams are those without tributaries. When a stream meets another stream of the same order its designation increases by one.

4. When two creeks join, take samples from each creek, but never immediately after their joining.

5. For the sampling position find a place where the speed of the current changes abruptly, from slow to fast to wash out the light fraction, or from fast to slow, where the heavy fraction accumulates. It is recommended to take a composite sample of both places. Examples of sampling sites are the meanders, small waterfalls, obstacles (natural or artificial) in the current, sandbanks on islands, etc. (Fig. 2).

Figure 2. A perfect spot for taking a representative HMC sample.

6. Sandy material containing small boulders should be preferred. Wash out the clay material before the actual panning of the sample (Fig. 3).

Figure 3. Eliminating the excess of clay from the sample.

7. Take a constant volume of sample at each sampling place. A 10 to15-liter sample is sufficient. Sieve the material twice. First with a sieve of 3-4 cm. Later with a sieve of less than 0.5 cm. Record an estimation of the volume of each of these macro-fractions. The fine material is then washed on a circular or square pan. The square pan is easier to use than the circular pan because you only need a back-and-forward movement. The circular pan requires a combination of circular and back-and-forward movements and is practical in shallow places. The selection of the pan model depends on the preferences of the user.

Figure 4. Panning a sample to obtain an HMC with a circular pan.

8. For regional exploration wash the sample down to a gray fraction. The gray fraction contains more light minerals than a black fraction. You can use the black fraction during local explorations.

Figure 5. Gray fraction of the HMC sample ALReady showing gold colors.

9. The process of panning has been described by many, but I have never known of anyone that learned to pan by reading a description. This is the kind of knowledge that needs to be transmitted by example. You can see how panning is done on YouTube (http://ow.ly/JOwS30eaPAK; ttp://ow.ly/8X6m30eaPFn; http://ow.ly/6KsM30eaPK3; and many others).

10. If the proper conditions exist, air-dry the samples at camp. If the original paper bag is in bad condition, put it inside a new paper bag, rather than re-bagging.

11. A preliminary description using a lens or a portable microscope (Fig. 6) can be done directly in the field (Valls, 1987). Send the sample to a specialized laboratory for a more detailed description.

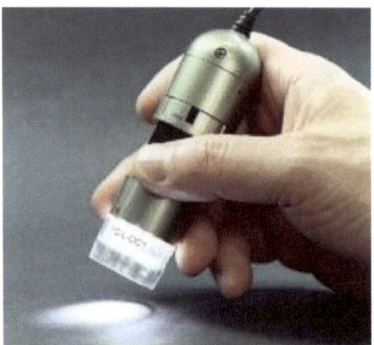

Figure 6. Manual microscope from Dino Lite.

12. Apart from the mineralogical description of each HMC sample, you can separate three fractions from each sample: the magnetic, the electromagnetic, and the heavy fraction. Obtain the magnetic fraction by using a magnetic separator (Fig. 7). You will need an electromagnet to separate the electromagnetic fraction. To obtain the heavy fraction use a heavy liquid such as bromoform, tetrabromoethane, tribromoethane, methylene iodide, or polytungstate. All these liquids are toxic and therefore you must do the separation under control conditions. After describing each of these fractions sent them for analysis. I recommended using Instrumental Neutron Activation Analysis (INAA) because it keeps the samples intact (http://ow.ly/3gkm30eaSIT).

Figure 7. The magnet used to separate the magnetic fraction of an HMC.

13. You can also collect by hand mono fractions of specific minerals (pyrite, zircon, etc.) and send them for analysis.

Density of samples

For a regional scale of 1:50 000 take 4-6 samples per square km from the first or second order streams. For country-size projects, you can use densities of 1 sample per square kilometer. Such grid was successfully used during a regional survey of the Amazon river in Brazil. Stream sediments at scales larger than 1:10 000 (approximately 10-12 samples per square km) are inadvisable. At such scale, soil sampling methods with or without a grid are more effective.

Soil sampling

When and where to use it

Use soil sampling to test anomalies previously defined by stream sediments or HMC surveys.

Sampling method

The sampling method depends in part on the type of analysis. The "normal" soil sampling consists of taking a 200-250g sample from an established soil horizon (geochemist prefer the top of the Horizon B) or at an established depth in a cloth or Kraft paper bag, following a grid. This can be a surveyed grid, a GPS-oriented grid or a "compass and length of step" grid.

Mobile Metal Ions (MMI) and Spatiotemporal Geochemical Hydrocarbons (SGH) are common methods, but here I will mention the Enzyme Selective Extraction method (ESE). It is one of the most discriminating of the selective analytical extractions in use today. ESE targets amorphous mixed oxide coatings. By selectively removing the amorphous manganese dioxide from these coatings, the mixed oxide coatings collapse, releasing trapped trace elements. Currently, the greatest depth of penetration of ESE for a mineral deposit is greater than 800 m (Fig. 8).

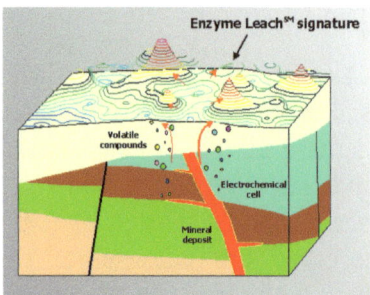

Figure 8. The formation of an enzyme selective extraction signature.

In the field, take a 200-250 g of material from the top of Horizon B and send it for preparation and analysis Alternatively, you can sieve it in the field using a -60 to -80 Mesh sieve. Collect a sample of 5-10 g from the top of Horizon B[1]. No other preparation is required.

Of the original 5-10 g, the laboratory will take a 0.75 g sample and leach it in an enzyme matrix containing a glucose oxidase solution at 30°C for 1 hour. The enzyme reacts with amorphous MnO_2 dissolving it. The metals are complexed with the gluconic acid present. The solutions are analyzed on a Perkin Elmer ELAN 6000, 6100 or 9000 ICP/MS. For each tray of 54 samples, the laboratory will add one blank, three duplicates, and 4 standards.

Currently, only Canadian Activation Laboratories Ltd. (www.actlabs.com) can do this analysis (code 7 or 7e). The method provides information on 66 elements, including gold with a 0.05 ppb detection limit.

Density of samples

The sampling grid should cover the investigated anomaly and extend beyond its limits. The ratio between profile and samples should represent the ratio of the target.

For example, if the target is a vein system, use a grid with larger spaces between profiles but shorter spaces between sampling points, e.g. 200x50m or 150 x 25m. A more "square" grid, e.g. 200

---

[1] Or from a constant depth.

x 100 will be more appropriate for an intrusive target.

Determining the best sampling method

Consider conducting experimental works prior to the sampling program to determine the best sampling and analyzing options. To do this, conduct a detailed sampling program across a known mineralization in or near the studied area. It is sufficient to have one or two detailed profiles perpendicular to the mineralized zone. These profiles must start and end well within the host unmineralized rock. You can see an example of such methodological work here- http://ow.ly/KGBF30ebdK2.

You must have a correct design of the survey and the correct method to define the anomaly. So, you need to answer the question- **What do I consider the most informative version of my test and how I will evaluate this?**

You should consider as the most informative version the one that brings the higher **contrast** of the information (the difference between the actual value and the "background" value in the population).

To select the best possible option, I recommend the use of the **Average Contrast Coefficient (ACC)** as defined by Soloviov A.P. and A. A. Matvieiev (1985). This coefficient is determined by the following equations:

$$\bar{X} = \frac{1}{n}\sum_{i=0}^{n} x_i$$

$$S = \sqrt[2]{\frac{\sum_{i=0}^{n}(x_i - \bar{x})^2}{n-1}}$$

$$X_{3+} = \bar{x} + 3S$$

$$\varepsilon = \sqrt[2]{\frac{X_{3+}}{\bar{x}}}$$

$$Y = \frac{1}{\log \varepsilon} \log \left(\frac{x_{max}}{\bar{x}}\right)$$

Where,

n- the amount of data
x_i- the individual value of the studied parameter
$\bar{x}$- the arithmetic mean (can be substituted by the median or absolute mode if the distribution law is unknown or if the number of samples is too small).
S- the standard deviation of the sample
x_{3+}- the lower value of the third level anomalies of the sample
x_{max}- the maximum value of the sample
Y - the average contrast coefficient of the studied parameter.

Chapter II
Establishing the quality of the data

All exploration programs must incorporate appropriate quality control techniques to minimize potential imprecision and bias in the data.

The best means of reducing both these errors are diligence and consistent adherence to protocols. Bias in quality is introduced through various mechanisms: unrepresentative sampling and contamination are the most common ones. These mechanisms can be dealt with by ensuring strict adherence to the sampling protocols. Unrepresentative sampling will be mitigated if sample collection techniques are thorough (Valls, 2016).

Contamination is a more complex problem to address. The major sources of contamination include:

1. Contamination by field staff during sample collection.
2. Contamination from the sampling device.
3. Contamination from the sample bag.
4. Contamination during sample processing such as from atmospheric deposition, wind, dust, etc.

Use quality control samples (QC) to identify where the contamination entered in the process. There are four main types of QC samples. Field blanks and duplicate samples detect contamination introduced throughout the sampling component of the program. Standard samples detect contamination introduced during the analytical process (the lab component of the program).

Allocate around 30% of the program budget to these QA&QC measures. All new programs should incorporate rigorous QA&QC until you can demonstrate a consistent, acceptable quality level of data.

The initial amount allocated depends on one or more of the following:

1. The level of experience of the field staff.
2. The type of sampling program. Impact assessment and survey (or baseline) studies require more QA&QC funds than compliance and regular sampling.
3. State of the sampling environment.

Here I present the abbreviated description, application, and protocol of the most important kinds of QC samples.

Duplicate samples

Duplicate samples are collected at one or more sites to assess the precision of the entire program (field and laboratory components). Duplicate measurements on a single sample (every 20-40th sample) yield the laboratory precision. The use of duplicates assumes that the variability among duplicates is affected by the sampling method or by the technician.

There are two other types of duplicates- **crushing preparation duplicate (CPD)** and **pulverization preparation duplicate (PPD)**. CPD are splits of coarse crushed material, or second sieved sample for soils, to measure the precision of laboratory preparation and analysis. Variation should be less than that of field duplicates as a correctly prepared sample is fairly homogenous. PPD is a duplicate taken from a pulverized sample to measure the precision of the pulverization and analysis. There should be minimal variation in these duplicates as the pulp should be homogeneous.

Blanks

Blanks are important in dealing with erratic results. Blanks may identify unsuspected contaminants associated with improper cleaning procedures, samples, shipping, sampling technique or air contaminants that may have been absorbed by the samples during collection. **Field blanks** should be exposed to the sampling environment at the sample site and handled in the same manner as the real sample. Consequently, they provide information on contamination resulting from the handling technique and from exposure to the atmosphere. A good use of **field blanks** will be during RC drilling or near the diamond saw at the core shake. For **field blanks,** a company could use rejects that have been ALReady tested to be barren or river sand from well-known barren zones or from limestones.

Laboratories routinely test for **equipment blanks** which are samples that have been used to clean sampling equipment. This type of blank is useful in documenting the effectiveness of the cleaning or decontamination of equipment. The results of these controls are available upon request and companies should request them to be included in their regular audits.

Standard samples

Standard samples are used to document the bias and precision of the analytical process. There are two types of standards. The choice as to what standard sample is selected depends on the expected concentrations being measured and whether comparable concentrations are available in existing standard samples.

The first and the simplest standard sample is an **uncertified standard sample**. In these case, a sample of many kilograms is collected by the Client and is pulverized and homogenized by a laboratory.

The second type of standard material is a **certified standard sample**. A certified standard sample is a portion of a large batch sample that was collected from one place at one time. The batch sample is preserved to ensure the stability of the certified variables and was subjected to analysis by a large number of independent laboratories using several different analytical techniques. A good provider of certified standards is CDN Laboratories in Canada, but you can find other providers online. The Company must use a different standard provider than the one its laboratory uses.

Laboratories will use certified standards for their own QC procedures. However, when implementing a sampling program, you must submit additional standard samples "blind" to the analyzing laboratory so that the reported value obtained under routine analytical conditions could be compared to the "true" value. You can submit non-blind standard samples to the laboratory, but these samples will receive special attention and will represent the best quality that the laboratory can produce.

External control

External control checks for the precision and reproducibility of the results of the main laboratory. The independent measurements on a single sample (every 100th sample) yield the main laboratory precision and reproducibility. The pulps are sent to another certified laboratory and submitted to the same analysis using the same type of technique. It is inadvisable to compare FA results with those obtained by an INAA technique for example.

Chapter III
Preparing the data

Cleaning the data

When zero doesn't mean it.

Before applying any exploration method, we should clean the data by first dealing with the below the detection limit (b.d.l.) values, substituting the missing data by a question mark (if we are planning to use Weka) or eliminating them all together, and then creating a classification class.

With b.d.l. values two of the most commonly used methods consist in changing the b.d.l. value by half of the value of the detection limit or by changing the b.d.l. value by the next significant digit (N.S.D.). Table 1 shows several examples.

Table 1. Treatment of b.d.l. values.

B.D.L.	Half	N.S.D.
>10.00	11.00	11.00
<5.00	2.50	4.00
<1.00	0.50	0.90
<0.10	0.05	0.09

Also, convert all units into one, such as ppm or ppb.

Different kinds of zeros

We always find below detection limits and/or zero values. Since most of the geological data respond to lognormal distributions, these "zero data" represent a mathematical challenge for the interpretation.

Amalgamation, e.g. adding Na_2O and K_2O, as total alkalis is a solution, but sometimes we need to differentiate between a sodic and a potassic alteration. Pre-classification into groups requires a good knowledge of the distribution of the data and of the geochemical characteristics of the groups that are sometimes unavailable. Considering the zero values equal to the limit of detection of the used equipment would generate spurious distributions in ternary diagrams. The same situation would occur if we replace the zero values by a small amount using nonparametric or parametric techniques (imputation).

The method that we are proposing takes into consideration the well-known relationships between some elements. For example, in copper porphyry deposits, copper values correlate well with molybdenum ones. While copper will always be above the limit of detection, many of the molybdenum values will be "rounded zeros". So, we take the real molybdenum values and establish a regression equation with copper, and then we estimate the "rounded" zero values of molybdenum by their corresponding copper values.

The method could be applied to any type of data, provided we first establish their correlation.

One of the main advantages of this method is that we obtain a value that depends on the value of another variable.

Are there any zeros in the house?

We start by recognizing that zero values exist in geology. For example, the amount of quartz in a foyaite (nepheline syenite) is zero since quartz and nepheline are incompatible (Trusova & Chernov, 1982). In binomial distributions, such as during the drilling of an ore body, you either intersect the ore body (1) or not (0). Another common zero is a North azimuth, although we can change that zero to the value of 360°. These are known as "essential zeros" (Aitchison, 2003) or "real zeros". They are irrelevant for as long as their population responds to a normal distribution.

In geochemistry, we also have "rounded zeros". In some cases, labs report below the detection limit (b.d.l.) values as zeros or nonexistent, while in most cases they just put the b.d.l. as the value for that parameter. These b.d.l. values are a similar problem to the "rounded zeros". Let me illustrate with the example proposed in Table 2.

Exploring Geological Data with Weka, CoDaPack, and iNZight

Table 2. Selection of the original data for the used example.

Au	Cu	As	Au	Cu	As	Au	Cu	As
0.3	330	13	0.2	270	14	0.2	320	21
0.2	80	116	0.1	90	32	0.5	920	53
0.1	40	23	0.1	270	9	0.2	180	46
0.2	190	19	0.3	820	18	0.4	1170	236
0.2	180	25	0.2	260	5	0.2	150	222
0.2	160	21	0.2	110	15	0.2	350	36
0.2	230	116	0.2	210	52	0.2	460	28
0.1	330	29	0.2	160	179	0.8	120	147
0.5	240	138	0.3	180	75	0.2	100	45
0.3	210	112	0.3	170	45	0.2	240	65
0.1	140	8	0.2	80	36	0.4	450	13
0.1	290	14	0.1	270	12	0.9	760	19
0.2	150	18	0.2	330	35	0.1	240	14
0.4	680	42	0.3	500	68	0.2	320	31
0.1	130	30	0.2	450	12	0.4	240	52
0.2	100	106	0.2	260	25	0.4	220	82
0.1	230	19	0.2	190	71	0.8	590	40
0.1	140	78	0.1	340	18	0.3	240	552
0.2	100	45	0.2	270	51	0.4	250	9
0.1	130	21	0.3	730	14	0.1	400	9
0.3	580	12	0.3	160	31	0.4	600	61
0.2	140	71	0.1	170	17	0.3	1940	101
0.1	220	17	0.1	440	12	0.2	130	43
0.1	120	49	0.2	300	9	0.3	250	6
0.4	120	307	0.09	20	5	0.2	340	18
0.1	160	20	0.1	240	7	0.1	130	24
0.09	150	14	0.1	170	31	0.2	180	118
0.09	150	13	0.1	520	9	0.3	150	55
0.1	290	18	0.4	360	139	0.1	170	13
0.1	140	33	0.2	180	58	0.2	110	29
0.2	420	10	0.2	400	8	0.3	240	138
0.2	190	15	0.1	840	79	0.1	180	120
0.2	170	8	0.2	180	61	0.5	140	42
0.1	70	12	0.1	150	20	0.3	190	110
0.2	310	11	0.2	280	83	0.3	240	9
0.2	160	14	0.2	280	84	0.4	310	19
0.3	220	37	0.8	380	149	0.3	400	64
0.1	90	10	0.4	210	56	0.1	410	12
0.1	130	17	0.5	810	33	0.2	220	39
0.3	110	16	0.1	160	12	0.1	240	13
0.1	190	12	0.1	220	11	0.1	1350	7
0.1	110	13	0.2	410	14	0.1	230	7
0.1	290	14	0.2	230	36	0.2	230	11
0.4	750	17	0.09	210	19	0.1	300	7
0.2	470	16	0.2	370	24	0.2	380	18
0.3	600	19	0.2	120	32	0.2	240	21
0.4	180	65	0.8	270	139	0.3	240	29
0.2	380	18	0.4	260	54	0.6	120	106
0.2	340	55	0.09	170	17	0.3	410	9

A ternary diagram can show these results (Fig. 9).

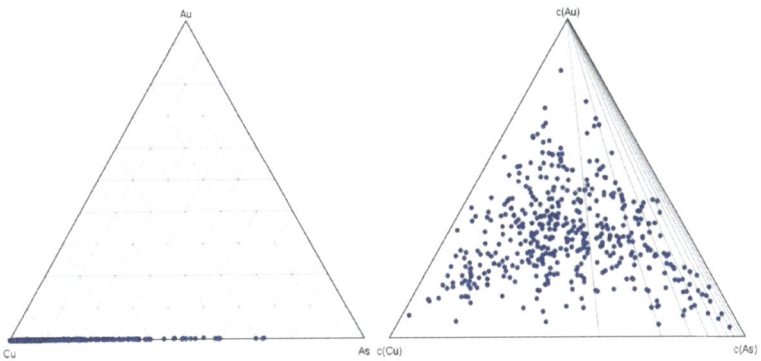

Figure 9. Ternary diagram of the studied data. To the left is the raw data, to the right is the same data centered using CoDaPack software.

Centering is incapable of completely solving the problem of the b.d.l. data, since part of them remains grouped along the Cu-Au axis.

21

Zero, zero... What shall I do with you?

Geologists- even those who are unknowledgeable of compositional data analysis- have been dealing with these problems for some time (Kashdan, Guskov, & Chimanskii, 1979). One of the most frequently used technique is an amalgamation (Aitchison, 1982). I ALReady discussed amalgamating and pre-classification as possible solutions, but both have serious limitations.

Considering the zero values equal to the limit of detection of the used equipment, or substituting it by some other constant (e.g. half the limit of detection) will generate spurious distributions, in ternary diagrams as seen in Fig. 9.

The same situation occur when we replace the zero values by a small amount (Bacon Shone & others, 2003) using nonparametric or parametric techniques (imputation). Even if we added the same small value to all the analyzed parameters, we would get the same spurious distribution.

How do I deal with spurious distributions?

Table 3 explains the procedure mentioned earlier that uses the well-known relationships between some elements to establish a regression equation, and then estimate the "rounded" zero values.

Table 3. Procedure for determining the b.d.l. values of molybdenum by using the values of copper.

All values			Real values		SUMMARY OUTPUT				Estimated Mo		
Au	Cu	Mo	Cu	Mo					Au	Cu	Mo
4.07	447.07	10.00	455.59	20.66	**Regression Statistics**				4.07	447.07	22.97
4.43	388.68	10.00	552.81	30.38	Multiple R	0.993656			4.43	388.68	17.75
3.80	413.46	10.00	582.62	33.36	R Square	0.987352			3.80	413.46	19.97
4.51	317.22	10.00	510.13	26.11	Adjusted R Square	0.986608			4.51	317.22	11.36
3.01	324.48	10.00	607.71	35.87	Standard Error	2.008715			3.01	324.48	12.01
2.81	280.80	10.00	705.21	45.62	Observations	19			2.81	280.80	8.11
4.44	280.00	10.00	250.00	10.00					4.44	280.00	8.04
4.23	255.00	10.00	857.08	60.81					4.23	255.00	5.80
3.60	250.00	10.00	714.87	46.59		Intercept	Cu		3.60	250.00	5.35
2.86	290.00	10.00	826.56	57.76	Coefficients	-16.9993	0.089408		2.86	290.00	8.93
1.57	300.00	10.00	536.82	28.78	Standard Error	1.678148	0.002454		1.57	300.00	9.82
2.67	455.59	20.66	845.11	59.61	t Stat	-10.1298	36.42962		2.67	455.59	23.73
4.25	552.81	30.38	844.20	59.52	P-value	1.28E-08	1.41E-17		4.25	552.81	32.43
4.61	582.62	33.36	876.55	62.76	Lower 95%	-20.5398	0.08423		4.61	582.62	20.66
2.60	510.13	26.11	828.78	57.98	Upper 95%	-13.4587	0.094586		2.60	510.13	26.11
4.05	607.71	35.87	832.87	58.39	Lower 95.0%	-20.5398	0.08423		4.05	607.71	28.78
3.97	705.21	45.62	664.17	41.52	Upper 95.0%	-13.4587	0.094586		3.97	705.21	30.38
1.09	250.00	10.00	731.04	48.20					1.09	250.00	33.36
4.78	857.08	60.81	270.00	10.00					4.78	857.08	35.87
3.07	714.87	46.59			ANOVA				3.07	714.87	41.52
4.05	826.56	57.76				df			4.05	826.56	45.62
1.44	536.82	28.78				1	17	18	1.44	536.82	46.59
3.47	845.11	59.61			SS	5354.833	68.5939	5423.426	3.47	845.11	48.20
3.04	844.20	59.52			MS	5354.833	4.034935		3.04	844.20	57.76
2.99	876.55	62.76			F	1327.117			2.99	876.55	57.98
2.58	828.78	57.98			Significance F	1.41E-17			2.58	828.78	58.39
1.90	832.87	58.39							1.90	832.87	59.52
0.69	664.17	41.52							0.69	664.17	59.61
0.83	731.04	48.20							0.83	731.04	60.81
0.09	270.00	10.00							0.09	270.00	62.76

According to Table 3, we can use regression equation (1) to estimate the b.d.l. values for Mo:
(1) Mo = -17 + 0.09*Cu
Figure 10 shows the result of this procedure.

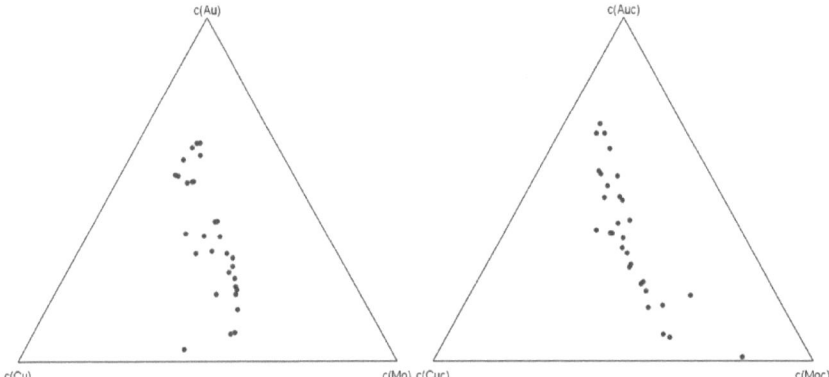

Figure 10. The procedure obtains a more linear distribution of the data (right) when compared to the original data (left).

How does CoDaPack deal with zeros?

J. A. Martín-Fernandez, J. Palarea-Albaladejo, & R. A. Olea (2011) explained the theory behind their algorithm to deal with zeros in CoDaPack. To use CoDaPack to replace b.d.l. values first make sure that such values are anteceded by a "<" or ">" symbol. See Table 4 as an example.

Table 4. An example of data with b.d.l. values.

Au	Cu	Mo	Au	Cu	Mo
<1	270.00	<10	3.47	845.11	59.61
4.51	317.22	<10	3.04	844.20	59.52
4.44	280.00	<10	1.90	832.87	58.39
4.43	388.68	<10	2.58	828.78	57.98
4.23	255.00	<10	4.05	826.56	57.76
4.07	447.07	<10	<1	731.04	48.20
3.80	413.46	<10	3.07	714.87	46.59
3.60	250.00	<10	3.97	705.21	45.62
3.01	324.48	<10	<1	664.17	41.52
2.86	290.00	<10	4.05	607.71	35.87
2.81	280.80	<10	4.61	582.62	33.36
1.57	300.00	<10	4.25	552.81	30.38
1.09	250.00	<10	1.44	536.82	28.78
2.99	876.55	62.76	2.60	510.13	26.11
4.78	857.08	60.81	2.67	455.59	20.66

To substitute these b.d.l. values, import your table in xls, xlsx, or csv format to CoDaPack (Fig. 11).

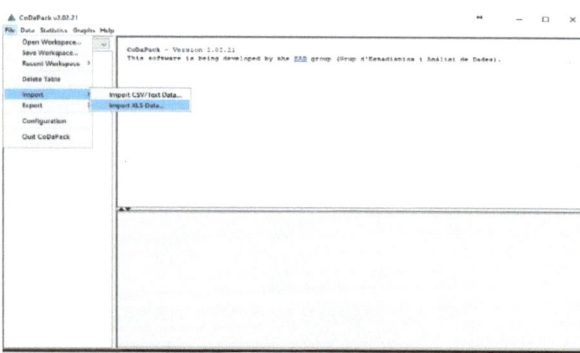

Figure 11. Import a file into CoDaPack

The b.d.l. data appears in the table with a specific format (Fig. 12).

Exploring Geological Data with Weka, CoDaPack, and iNZight

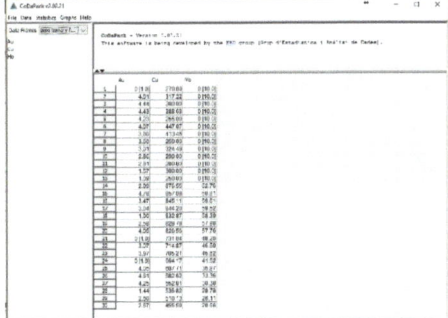

Figure 12. B.d.l. values in CoDaPack.

Next, select the rounded zero replacement option from the Data tab (Fig. 13).

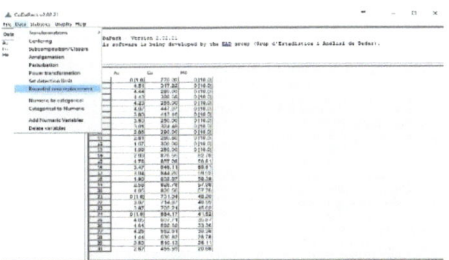

Figure 13. Rounded zero replacement algorithm from CoDaPack.

Select all the elements, including the b.d.l. values. Unselect the "Closure result" option and adjust the detection limit proportion (DL) if necessary (Fig. 14).

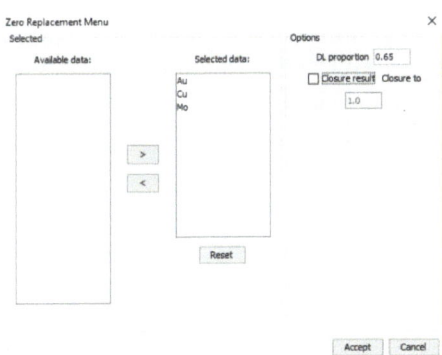

Figure 14. Zero replacement menu on CoDaPack.

CoDaPack converts the b.d.l. values (Fig. 15).

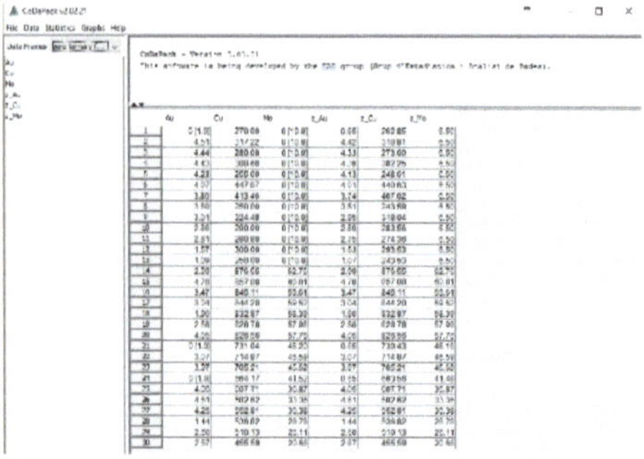

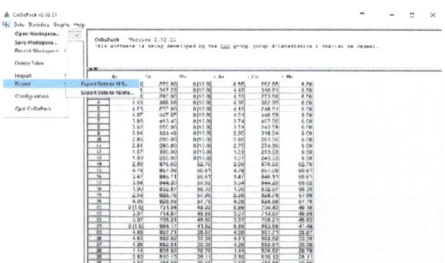

Figure 15. Data without b.d.l. values.

Now, export the transformed data back to an xls format or to R for further processing (Fig. 16).

Figure 16. Exporting the data without b.d.l. values to Excel.

Amalgamation

Besides the added advantage of eliminating the rounded zero effect, the amalgamation or addition of elements has other uses.

I will use the data represented in Table 5 as an example. There I amalgamated the data from SiO_2, Al_2O_3, and the alkali elements (Na_2O and K_2O). The amalgamation can be done within CoDaPack (Fig. 17) but it is easier to do it directly in Excel.

Exploring Geological Data with Weka, CoDaPack, and iNZight

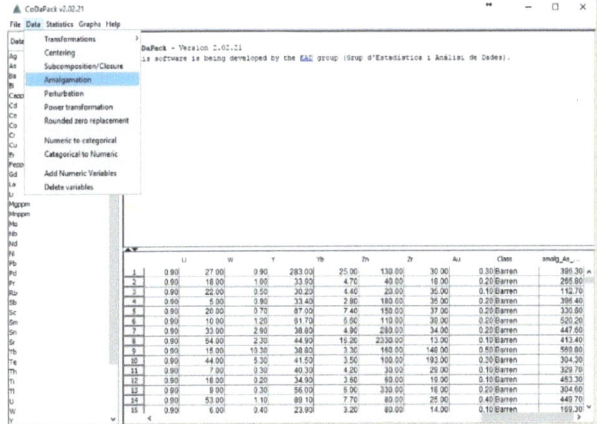

Figure 17. Amalgamation in CoDaPack.

Table 5. Example of amalgamated petrologic data.

SiO₂	Al₂O₃	SiO₂+Al₂O₃	Fe₂O₃	CaO	MgO	MnO	K₂O	Na₂O	K₂O+Na₂O	P₂O₅	TiO₂	LOI
72.11	13.80	85.91	2.73	1.90	0.68	0.06	3.07	4.32	7.39	0.08	0.26	0.99
70.60	15.75	86.35	2.51	4.05	0.79	0.01	0.38	4.70	5.08	0.13	0.40	0.68
35.54	35.04	70.58	10.08	0.90	0.92	1.17	3.53	2.85	6.37	3.63	2.55	3.79
30.40	39.44	69.83	12.19	0.80	0.81	0.72	3.74	3.26	7.00	2.79	3.17	2.68
38.81	28.40	67.21	14.98	1.09	1.04	0.53	3.62	2.53	6.16	2.57	3.87	2.56
42.46	12.13	54.59	24.96	1.17	1.18	0.61	3.08	3.25	6.33	3.55	4.10	3.50
34.19	31.18	65.37	13.41	1.09	1.11	0.89	2.62	4.00	6.62	3.91	4.06	3.58
37.29	34.21	71.50	10.09	0.90	0.77	0.89	2.64	4.14	6.77	3.88	2.64	2.68
25.22	36.59	61.81	19.68	0.62	1.07	0.92	3.06	3.12	6.18	3.05	3.91	2.80
29.65	45.11	74.76	12.35	0.50	0.63	1.14	0.32	0.25	0.57	2.58	3.87	3.60
35.17	33.47	68.64	11.66	0.51	1.14	1.17	4.18	2.60	6.77	3.45	3.53	3.08
33.13	36.05	69.17	14.27	0.58	1.03	0.56	3.27	2.80	6.07	2.99	2.76	3.40
33.00	32.56	65.57	14.78	0.59	1.08	0.51	4.20	2.85	7.05	3.78	3.71	2.96
57.86	16.20	74.06	6.75	5.07	2.61	0.27	3.79	1.64	5.43	0.62	1.03	4.16
60.97	17.35	78.32	4.87	4.76	3.28	0.08	1.40	4.11	5.51	0.24	0.83	2.11
66.93	16.50	83.43	2.80	2.10	0.59	0.08	4.05	5.50	9.55	0.10	0.25	1.10
71.18	15.20	86.38	2.45	3.59	0.81	0.02	0.42	4.80	5.22	0.15	0.58	0.80
60.79	16.80	77.59	5.36	4.89	2.60	0.09	2.89	1.56	4.45	0.56	0.98	3.48
60.57	17.56	78.13	4.58	4.68	3.56	0.15	1.39	4.09	5.48	0.58	0.75	2.09

Let us see two cases where amalgamation adds more information to the interpretation of the data (Martín Fernández et al., 2003).

In Figure 18, you see the histograms from SiO₂ and Al₂O₃. Although we have two different types of rocks in the data set, it is difficult to see the difference between them. But if we plot the histogram of the amalgamated values of silica and aluminum (Fig. 19), then the presence of the two types of rocks is evident.

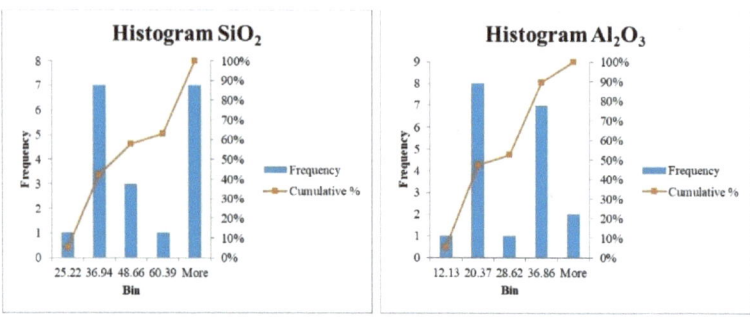

Figure 18. Histograms for silica and aluminum.

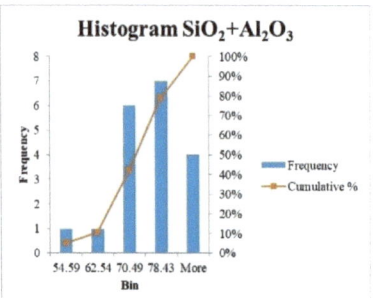

Figure 19. Histogram of the amalgamated values of silica and aluminum showing the presence of two types of rocks.

A ternary diagram of amalgamated values is also informative. Figure 20 shows a ternary diagram of SiO_2, Al_2O_3, and Fe_2O_3 on the left side, while on the right side we have a ternary diagram with almost twice more information (5 vs 3 components) by amalgamating the silica and the aluminum and the alkali elements.

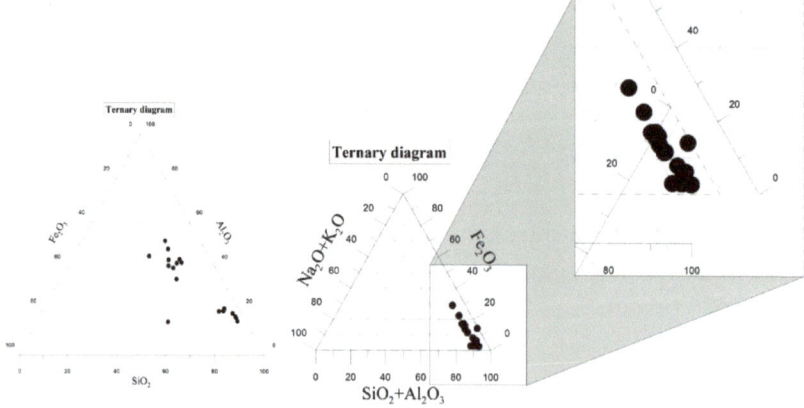

Figure 20. Ternary amalgamated diagrams.

Another example of amalgamation is the Rare Earth Elements (REE). All REE (Fig. 21) contain the same information so it will simplify further processing, if we add them into one column (REE).

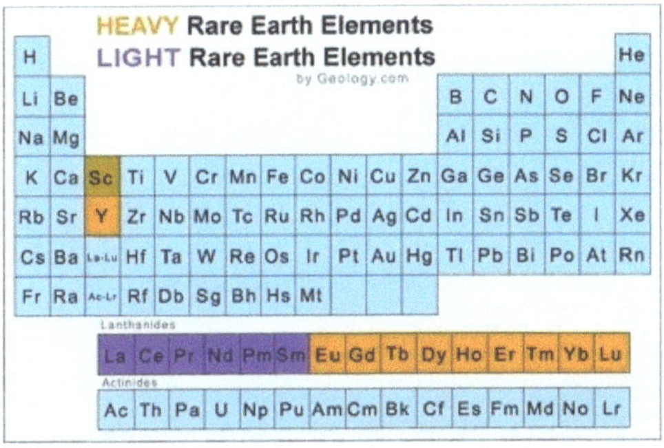

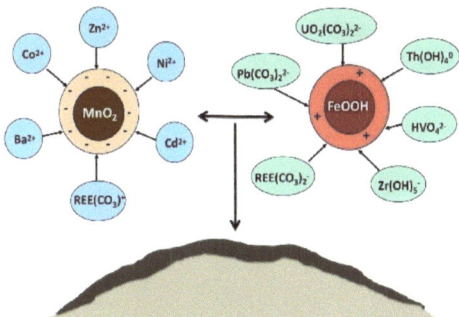

Figure 21. List of REE to combine in just one column.

Avoid amalgamating the REE if you are conducting detailed petrological studies, where you will need to compare the behavior of some of these elements (Petrelli, Poli, Perugini, & Peccerillo, 2005).

Normalization by Fe and Mn

To eliminate the reducing effect of Fe and Mn (Fig. 22), normalize the data as shown in Table 6.

Figure 22. Simplified electrochemical model for the sorption of trace metal species in seawater on the charged surfaces of colloidal or particulate Mn oxide and Fe oxyhydroxide, from Journal Elements, June 2017, Volume 13, Number 3.

Table 6. Elements to normalize.

Elements	Normalise with
Ba, Co, Zn, Ni, Cd	Mn
Pb, U, Th, V, Zr	Fe
REE	Mn+Fe

Estimating the erosional level

The basis for this technique developed by Soviet geochemists is that certain elements tend to concentrate on top of the ore body ("supra-elements") and others below it ("infra- elements"). The erosional level is estimated by dividing standardized values of the supra- elements over the infra- elements. The higher the obtained value, the further up is the current erosional level from the ore body, and therefore, anomalies in such areas are more promising.

There are ways to determine which specific elements are supra or infra in your target, but you can safely assume that Ag, Hg, Ba, and Zn will always be supra, while Ni, Co, and Cr will be infra-elements.

Use Excel to standardize your data using the function *standardize(X,mean,stdev)* and then estimate the value of the erosional level (Fig. 23).

ESTE	NORTE	Ba	Ag	Ni	Co	Ba, n	Ag, n	Ni, n	Co, n	Sample	Erosional level
421220	588931	370.00	1.00	4.30	9.50	0.30	-0.12	-0.83	-0.88	MQ0547	-1
419932	588333	30.00	0.04	1.00	20.00	-1.68	-0.40	-0.88	-0.32	MQ4070	1
416267	587977	250.00	10.00	2.70	9.90	-0.40	2.46	-0.85	-0.86	MQ2902	-2
417910	587293	320.00	0.17	51.60	22.20	0.01	-0.36	-0.04	-0.20	MQ2904	1
417873	587264	180.00	0.04	115.00	41.30	-0.81	-0.40	1.01	0.81	MQ2905	-1
417781	587173	340.00	0.04	109.00	42.70	0.12	-0.40	0.91	0.89	MQ2903	-1
420753	586982	590.00	0.07	4.30	5.60	1.58	-0.39	-0.83	-1.09	MQ4052	-1
421652	586937	470.00	0.02	144.50	57.10	0.88	-0.40	1.50	1.65	SL0822	0
Mean		318.75	1.42	54.05	26.04						
Stdev		171.92	3.48	60.15	18.82						

Figure 23. An example of an erosional level estimation.

Use the function *INT(x)* to simplify the results. Areas of big negative values of the erosional level are less promising, regardless of the contrast of the anomalies they host.

Class

For classification purposes in Weka, we need to have a column at the end, defined as a class. It could contain text or numbers, but avoid having the same information as a parameter. For example, if you are classifying the data set by gold results, unselect the gold column while doing classifications.

To create the classes, determine intervals of value for the element (e.g. Q2, Q3, and Q4) and use a conditional (if) to create classes as barren, low, medium, and high grade. Or you can define only two categories, e.g. "Interesting" above Q2 and "Barren" below Q2.

What is Weka?

Weka contains a collection of visualization tools and algorithms for data analysis and predictive modeling, with graphical user interfaces for easy access to these functions (Fig. 24).

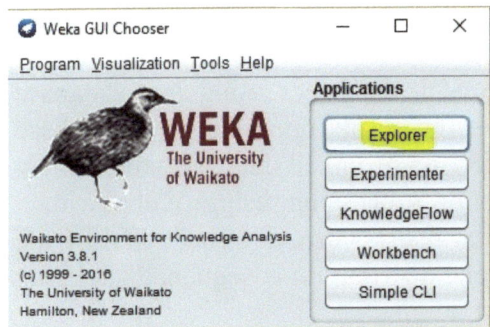

Figure 24. The interface of Weka.

The original non-Java version of Weka was a Tcl/Tk frontend to mostly third-party modeling algorithms implemented in other programming languages, plus data preprocessing utilities in C, and a Makefile-based system for running machine learning experiments. This original version was primarily designed as a tool for analyzing data from agricultural domains, but the more recent fully Java-based version (Weka 3), for which development started in 1997, is now used in many different application areas, for

educational purposes and research. Advantages of Weka include:
1. Free availability under the GNU General Public License.
2. Portability, since it is fully implemented in the Java programming language and thus runs on almost any modern computing platform.
3. A comprehensive collection of data preprocessing and modeling techniques.
4. Ease of use due to its graphical user interfaces (Frank, Hall, & Witten, 2016).

It is impossible to explain here the Weka algorithms and procedures. You can complete an online course at http://ow.ly/twHn30drfY7.

Description

Weka supports several standard data mining tasks like data preprocessing, clustering, classification, regression, visualization, and feature selection. All of Weka's techniques are predicated on the assumption that the data is available as one flat file or relation, where each data point is described by a fixed number of attributes. Data points can be numeric or categorical, but it supports other attribute types.

Weka provides access to SQL databases using Java database connectivity and can process the result returned by a database query. You can convert a collection of linked database tables into a single table that is suitable for processing using Weka with a separate software. Another area that is currently uncovered by the algorithms included in the Weka distribution is sequence modeling (Witten & Frank, 2005).

User Interfaces

Weka's main user interface is the Explorer, but you can access the same functionality through the component-based Knowledge Flow interface and from the Command Line. The Experimenter allows the systematic comparison of the predictive performance of Weka's machine learning algorithms on a collection of datasets.

The Explorer interface features several panels providing access to the main components of the workbench.

The Preprocess panel has facilities for importing data from a database, a comma-separated value (CSV) file, etc., and for preprocessing this data using filtering algorithms. Use these filters to transform the data (e.g., turning numeric attributes into discrete ones) and to make it possible to delete instances and attributes according to specific criteria.

The Classify panel enables applying classification and regression algorithms to the resulting data set, to estimate the confidence of the resulting predictive model, and to visualize erroneous predictions, receiver operating characteristic (ROC) curves, etc., or the model itself (if the model is amenable to visualization like, e.g., a decision tree).

The Associate panel provides access to association rule learners that attempt to identify all interrelationships between attributes in the data.

The Cluster panel gives access to the clustering techniques in Weka, e.g., the simple k-means algorithm. Weka can learn a mixture of normal distributions through the implementation of the expectation-maximization algorithm.

The Select attributes panel provides algorithms for identifying the most predictive attributes in a dataset.

The Visualize panel shows a scatter plot matrix, where individual scatter plots can be selected and enlarged and analyzed further using various selection operators (Frank et al., 2016).

Extension packages

In version 3.7.2, a package manager was added to allow the easier installation of extension packages. Some functionality has since moved into such extension packages, but this change also makes it easier for other to contribute extensions to Weka and to maintain the software. This modular architecture allows independent updates of the Weka core and individual extensions.

History

Following is a summary of the development history of the Weka software.
1. In 1993, the University of Waikato in New Zealand began development of the original version of Weka, which

became a mix of Tcl/Tk, C, and Makefiles.
2. In 1997, the decision was made to redevelop Weka from scratch in Java, including implementations of modeling algorithms.
3. In 2005, Weka received the SIGKDD Data Mining and Knowledge Discovery Service Award.
4. In 2006, Pentaho Corporation acquired an exclusive license to use Weka for business intelligence. It forms the data mining and predictive analytics component of the Pentaho business intelligence suite.
5. All-time ranking on Sourceforge.net as of 2011-08-26, 243 (with 2,487,213 downloads) (Frank et al., 1999).

Preparing the Data with Weka

Reducing the size of the dataset

Many of the parameters in your data set contain useless or redundant information. This is especially true for multi-element geochemical analysis. I will present a group of techniques to reduce the size of the dataset below. Follow the numbered instructions.

Wrapper

1. Run Weka.
2. Select Explorer from the Weka interface.
3. Open file (Fig. 25).

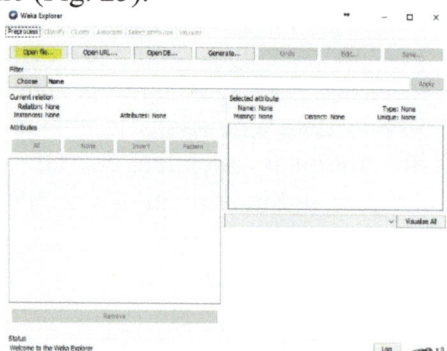

Figure 25. Open file from Explorer in Weka.

4. Go to Select attributes.

5. Attribute evaluator - WrapperSubsetEval (Fig. 26).

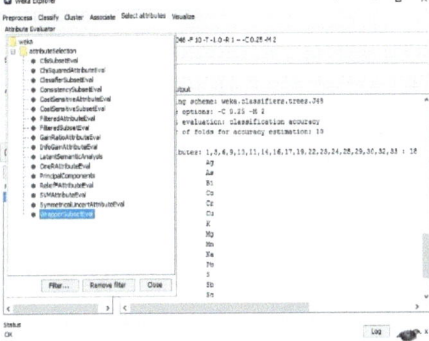

Figure 26. Select the attribute evaluator.

6. Left click (LC) on wrapper.
7. Classifier - select J48/Folds=10/Seed=1/Threshold= -1 (Fig 27).

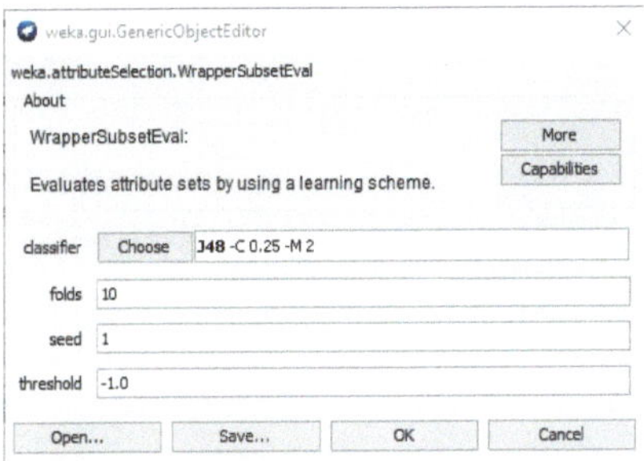

Figure 27. Parameters of the wrapper.

8. Search method- BestFirst/LC/direction: backwards, OK (Fig. 28).

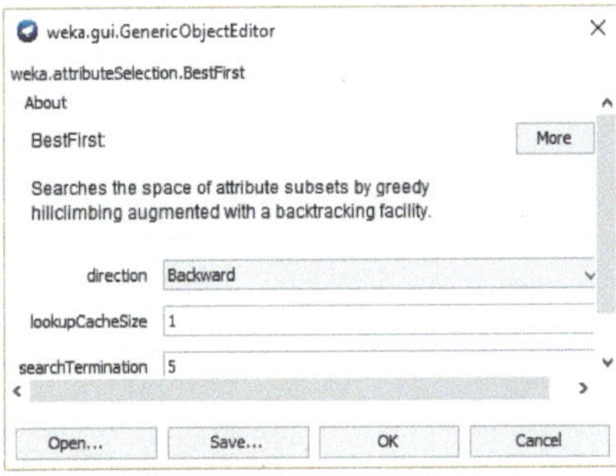

Figure 28. Parameters of the BestFirst.

9. Use full training set. In a second pass, you can also try

Cross-validation 10.
   10. Start.
This is a good but slow method. For the analyzed data set, Weka selected 18 elements of 35 as the most informative elements with a confidence of 92%, as shown in the line "Merit of best subset" (Fig. 29).

Figure 29. The wrapper method selected the 16 most informative elements of the data set.

Eliminate Redundant Information

1. Select Explorer from the Weka interface.
2. Open file.
3. Classify/Meta/AttributeSelectedClassifier/ LC.
4. Classifier: Naïve Bayes (NB).
5. Evaluator: Wrapper/LC/NB/Threshold= -1/OK (Fig. 30).
6. Search= BestFirst/LC/Backwards/OK/OK.

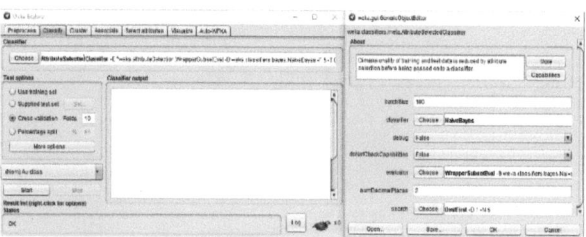

Figure 30. Parameters of the Weka classifier.
6. Cross-validation = 10.
7. Start.

For the analyzed dataset, Weka suggests the elimination of Ca, Fe, K, La, Mn, P, Sr, Th, Tl, U, V, and Bi with a 92% confidence (Fig. 31). Please note that Bi was selected as informative, so it should be kept.

Exploring Geological Data with Weka, CoDaPack, and iNZight

```
Selected attributes: 6,7,12,14,15,17,21,26,27,29,30,31 : 12
                     Bi
                     Ca
                     Fe
                     K
                     La
                     Mn
                     P
                     Sr
                     Th
                     Tl
                     U
                     V

=== Stratified cross-validation ===
=== Summary ===

Correctly Classified Instances     5942        91.5139 %
Incorrectly Classified Instances    551         8.4861 %
Kappa statistic                       0.5689
Mean absolute error                   0.0637
Root mean squared error               0.1832
Relative absolute error              54.4071 %
Root relative squared error          75.774  %
Total Number of Instances          6493
```

Figure 31. Redundant elements from the dataset.

Faster Method

1. Open file.
2. Classify/Meta/AttributeSelectedClassifier/ LC.
3. Classifier: NB (or J48).
4. Evaluator: CfsSubsetEval or wrapper/OK (Fig. 32).

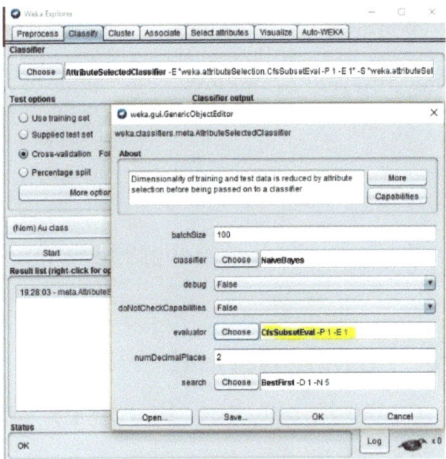

Figure 32. Faster method for selecting the most informative elements.

5. Start.

This method is fast and suggests the following elements as the best with a 92% of confidence: As, Bi, Cu, Ga, Ti.

When using instead J48 as the classifier and the wrapper as the evaluator, Weka would suggests the same elements with a 92% accuracy but it will be much slower.

Even Faster Method

This method ranks each parameter in order to selects relevant elements, but it fails to detect redundant ones.
1. Open file.
2. Classify/Meta/AttributeselectedClaassifier/LC.
3. Classifier: NB.
4. Evaluate: GainRatioAttributeEval.
5. Search: Ranker.
6. The number to select: -1 will rank all elements, but you can specify the number of elements, e.g. 7 will rank the first seven elements.

Weka, with a confidence of 86.8% selected: Bi, Cu, Mn, Ag, As, S, Fe, and Ca as the most relevant. If you use the test set, Weka identifies the same elements but with a confidence of 93.1%.

You can also try other types of algorithms. Table 7 shows the results for the different types I tested.

Table 7. Results using different evaluators.

Bi	Cu	Mn	Ag	As	S	Fe	Ca	Evaluate	Search	Confidence	Amount
1	1	1	1	1	1	1	1	1 GainRatioAttributeEval	N	93.1	8
1	1	1	1	1	1	1	1	1 SymetricalUncertattributeEval	N	93.1	8
1	1	1	1	1	1	1	1	1 InfoGain	N	93.1	8
1	1	1	1	1	1			OneRAttributeEval	N	93.2	6
1	1			1	1			Relief	N	94.0	4

Relief and OneRAttribute Eval are less recommended.

An Even Faster Faster Method

An even faster method, but one that does not estimate the confidence, follows:
1. Open file.
2. Go to Select attributes tab.
3. Evaluator- choose GairRatioAttributeEval.
4. Search- choose Ranker.
5. Use full training set.
6. Start.

Weka will give you a ranked list of all elements, so you can decide, for example, to keep only the first few ones on the list. In this case, the first 8 elements recommended are- Bi, Cu, Mn, Ag, As, S, Fe, and Ca. You can also test different evaluators, as it was shown in Table 7.

Another evaluator is PC that gives Principal Components associations of elements.

Once you have determined the elements that you are going to work with, remove the rest from the data set before continuing the study. You can save them as separated files.

Preparing a Training and a Testing Set

For many Weka applications, the program will use a training set created from the original data set. In general, you will get better results if you do that, because the test data is part of the original data set.

While the results will be sometimes significantly worse, I recommend you divide the original dataset in two: the training set that will be used to explore the properties of the original data set and a testing set to get the best results from the training set. The testing set should be representative of the training set and should exclude any common data.

It is possible to do this using Excel, but Weka allows us to obtain these sets easily. Just follow these steps:
1. Open your reduced file.
2. Go to Preprocess/Filter/Unsupervised/Instance/Resample.
3. LC on Resample.
4. Select False/True/1/30/OK/Apply.
5. Save as Training set.

Chapter IV
Exploring the data

Exploratory data analysis with Weka

By opening a dataset, you can see the basic statistics on the right side of the panel (Fig. 33).

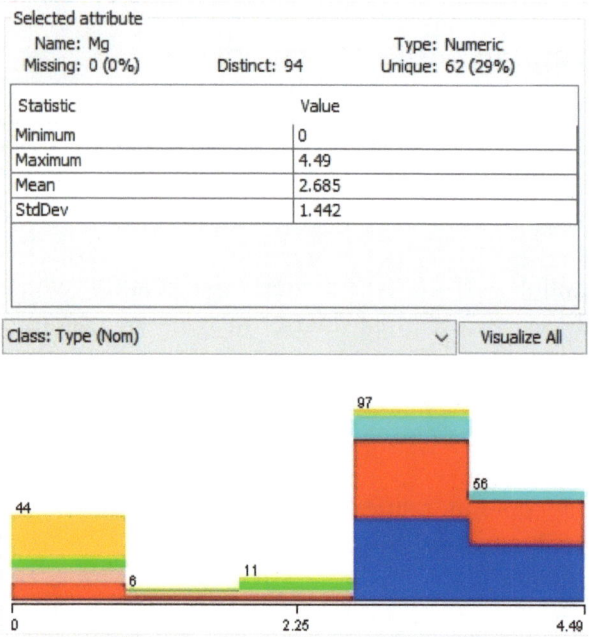

Figure 33. Statistical parameters of an attribute.

Histograms

A more useful method is to see the histograms of each attribute. To obtain these histograms:
1. Open file.
2. Choose Filter/Unsupervised/Attribute/Discretize.
3. LC/ Set the number of bins between 10-15/OK/Apply (Fig. 34).

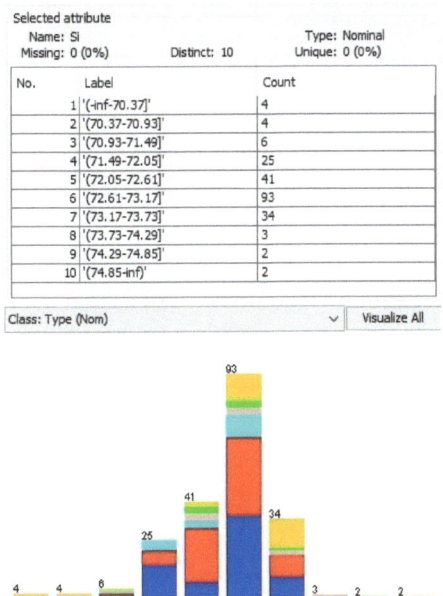

Figure 34. Histograms and bins for the selected attribute.

You can see all histograms at the same time by selecting Visualize All (Fig. 35).

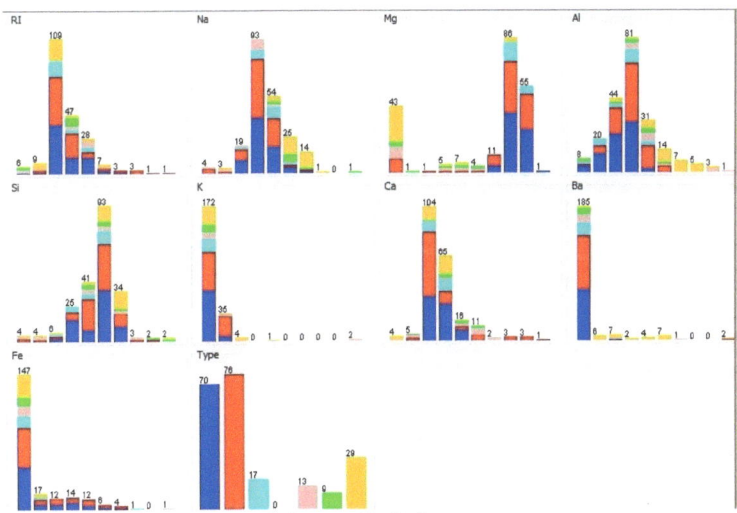

Figure 35. All histograms of the studied dataset.

Some classifiers, like Naïve Bayes, work better with discretized data. For the rest, click on Undo after you obtained the histograms.

Removal of statistical outliers

1. Open file.
2. Filter/Unsupervised/Attribute/InterQuartileRange/Apply.
3. Filter/Unsupervised/Instance/RemoveWithValues/LC.
4. Nominal index- last/OK/Apply.
5. Save your new table without outliers.

Clustering

Define number of clusters

1. Open file.
2. If your class is categorical, remove it or use numbers if possible.
3. Cluster tab.
4. Select XMeans.
5. Start (Fig. 36).

```
Time taken to build model (full training data) : 0.48 seconds

=== Model and evaluation on training set ===

Clustered Instances

0      4765 ( 73%)
1       915 ( 14%)
2       390 (  6%)
3       423 (  7%)
```

Figure 36. Determining the number of clusters.

EM also determines the number of clusters and you can use the class if it is categorical, but it is slow.

1. Open file (reduced).
2. Cluster tab.
3. Select EM.
4. Select training set reduced/Close.
5. Start (Fig. 37).

```
Time taken to build model (full training data) : 824.96 seconds

=== Model and evaluation on training set ===

Clustered Instances

0       675 ( 10%)
1       948 ( 15%)
2       126 (  2%)
3       353 (  5%)
4       283 (  4%)
5      2752 ( 42%)
6       160 (  2%)
7       268 (  4%)
8       155 (  2%)
9       773 ( 12%).
```

Figure 37. Clusters determined by the EM method.

Clusters 5, 1, 9, and 0 are the most significant. So, we also obtained 4 clusters by this method.

When (or once) you know the number of clusters

Once you know the number of clusters, you can use KMeans
1. Open file (reduced and without outliers).
2. Go to preprocess tab.
3. Filter/unsupervised/attribute/AddCluster.
4. Cluster tab.
5. KMeans/LC.
6. Input the number of clusters.
7. Select the class to cluster.
8. Start .
9. Go to Classify tab.
10. Classify via clusters (left side of the screen)/LC/KMeans/ # of Clusters.
11. Cross-validation.
12. Start.

Only two clusters

A useful technique to quickly explore your data is to force a cluster analysis with two groups. Accepting the concept that you have a mineralized zone in the limits of your license and that such mineralized zone will be different from the barren areas, a 2-component cluster should divide your data in a large group (barren) and a smaller group (potentially mineralized).

Plotting the coordinates of the smaller group on a map will give you an idea of the ore potential of your target (Fig. 38).

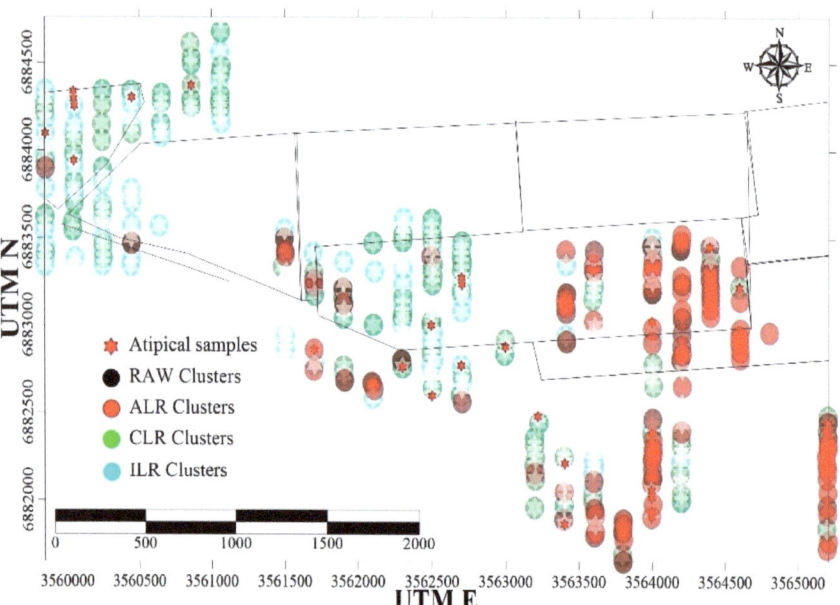

Figure 38. An example of the use of 2-component clusters to identify potential ore zones.

Based on these results, the Client arranged the limits of its claims to include the newly identified targets to the south of their current licenses.

To obtain the actual clusters, I recommend using this Excel macro (http://ow.ly/lopq30dBkkj) or CoDaPack.

Chapter V
Using the experimenter in Weka

While I believe that the Experimenter should be run as a first pass to test different classification models, I decided to explain the individual methods first. The Experimenter will tell you the general effectiveness of the different methods you want to test, so later you can do more detailed modeling in the Explorer mode.

To use the Experimenter:
1. Run Weka.
2. Select the Experimenter.
3. Select New (Fig. 39).

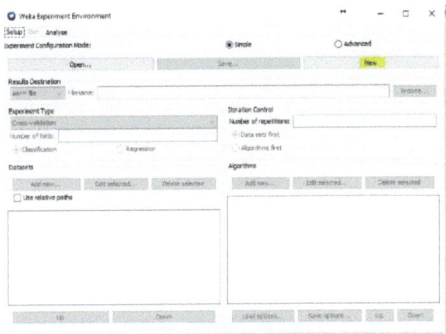

Figure 39. Select a new experiment option.

4. Left side add new folder.
5. Right side add new classifiers (Fig. 40).

Figure 40. The menu of the Weka Experiment Environment.

I suggest you test the following classifiers:

5a. ZeroR to establish the baseline. Any classifier must be higher than this value.

5b. OneR if you have a single classifier in your data, this will give you the best value.

5c. Naïve Bayes if the samples are truly independent and all are equal, Naïve Bayes will give you the best result.

5d. Lazy iBk Test for different K values (e.g. 1, 5, 10, etc). If all attributes are equal, this will give you the best result.

5e. J48 is the most efficient classifier and has the choice to create classification trees.

6. Go to the Run tab and click Start.
7. Go to the Analyze tab, click Experiment, click Show standard deviation and click Perform test (Fig. 41).

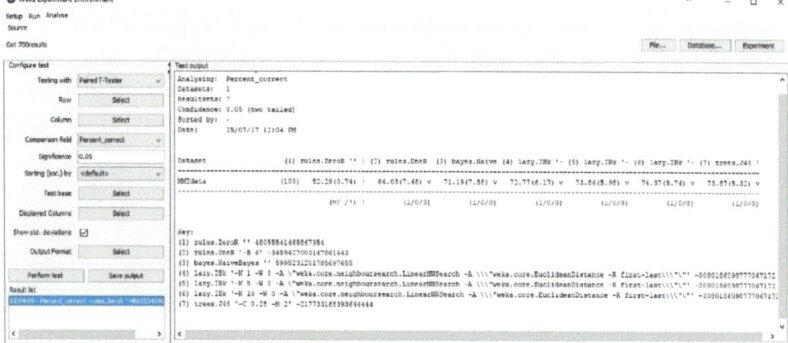

Figure 41. Results of the different classifiers.

You can see that all the classifiers performed significantly better than ZeroR and OneR, so J48 is the recommended classifier.

8. If you want to save the details of the experiment, go back to Setup and select a type of file, a name and a place where to save, click on the Run tab, and click on Start (Fig. 42).

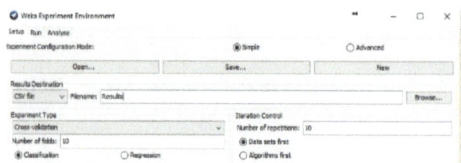

Figure 42. Saving the results of your experiment.

Chapter VI
Cost estimation with Weka

Evaluating the cost of your prediction

Weka finds the best qualifier and estimates its cost. In some cases, a classifier with a more modest performance could be significantly cheaper than a more efficient classifier. Cost analysis is used to estimate the cost of the errors.

Let start with a simple confusion matrix (Fig. 43).

	A	B
Ho	True Positive (TP)	False Negative (FN)
H1	False Positive (FP)	True Negative (TN)

	A	B
Ho	A	A/B
H1	B/A	B

Cost of errors = B/A + A/B

Value of the method = A + B

Figure 43. Simple confusion matrix.

At the beginning of an exploration program it is better to say that something is perspective when it is not (FP) than to say that something is not perspective when it is (FN). In this case, I suggest giving the FN a value higher than 1 (e.g. 2 or 3).

At an advanced stage of exploration, the reverse is true (and the cost of the error is greater), so the value of the FP should be higher than 5 (e.g. 5-10).

A more complex matrix is shown in Fig. 44.

	A	B	C	D
Ho	TPa=a	FPb/a	FNa/c	FNd
H1	FPb/a	TPb=b	FNb/c	FNb/d
H3	FPc/a	FPc/b	TNc=c	FNc/d
H4	FPd/a	FPd/a	FNd/c	TNd=d

	A	B	C	D
Ho	A	A	A	A
H1	B	B	B	B
H3	C	C	C	C
H4	D	D	D	D

Cost of errors = FPd/a+FPc/b+FNb/c+FNd+FPb/a+FPc/a+FPb/a+FPd/a+FPd/a+FNa/c+FNa/c+FNd/c+FNb/d+FNc/d

Value of the method = A + B + C + D

Figure 44. A complex matrix of confusion.

The objective of the test is to find the classification method with the smallest cost, independently of the accuracy.

1. Open file on Weka Explorer.
2. Go to Classify tab.
3. Filter/Meta/CostSensitiveClassifiers/LC/.
4. MinimizeExpectedCost= TRUE (experiment with FALSE).
5. Set cost matrix (see Fig. 45).
6. Select the classifier.
7. Start.
8. Repeat with a different classifier.

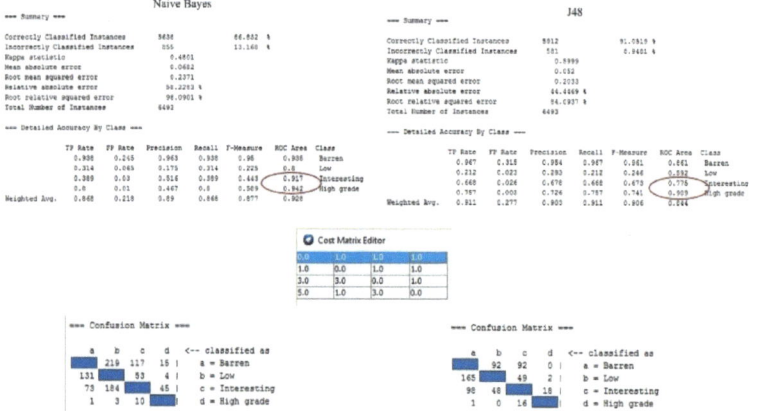

Figure 45. Cost analysis.

You can evaluate the cost directly in Weka, but I find it easier to test the different options in Excel (Table 8).

Table 8. Comparing the cost of two classifiers with Excel.

	Naïve Bayes	J48	Cost (time)	Naïve Bayes	J48
Total correctly classified	87%	91%			
Incorrect high grade	14	16	5	70	80
Incorrect interesting	302	164	3	906	492
Incorrect low	188	216	2	376	432
Incorrect barren	351	184	1	351	184
				1703	1188

I compiled the data from Table 8 after testing two classifiers (Fig. 45) by adding the number of incorrect determinations for each class in the confusion matrix. Table 8 shows that while J48 is more efficient as a classifier in general, if you were interested in reducing the cost of the error of misidentifying high-grade samples, then Naïve Bayes will be the best option.

ROC

Another useful method compares the accuracy of multiple classifiers by their receiver operating characteristics (ROC). You achieve this by using the Knowledge Flow section of Weka.
1. Add an Arff loader.
2. Go to the Evaluation tab.
3. Add a class assigner / Class Value Picker / and a Cross Validation.
4. Go to the Classifier tab and add all the classifiers you want to test.
5. Go to the Evaluator tab.
6. For each classifier add a ClassifierPerformanceEvaluator.
7. Go to the Visualization tab.
8. Add a ModelPerformanceChart.
9. RC on each of the first three icons and select data set to link them.
10. RC on the Cross-validation fold maker and select training set and later test set to join each of the classifiers.
11. RC on each of the classifiers and select the batch classifier to link to the ClassifierPerformanceEvaluator.
12. RC on the ClassifierPerformanceEvaluator and select threshold data to link to the ModelPerformancee chat.
13. RC on the Arff loader/Configure/ select data set.
14. RC on the ClassAssigner/Configure/select the class.
15. RC on the ClassValuePicker/Choose the parameter to test.
16. RC on Arff loader again/Start loading and see check for errors on the lower panel of Weka (Fig. 46).
17. Now RC on the ModelPerformanceChart on the right and select show cart (Fig. 47).

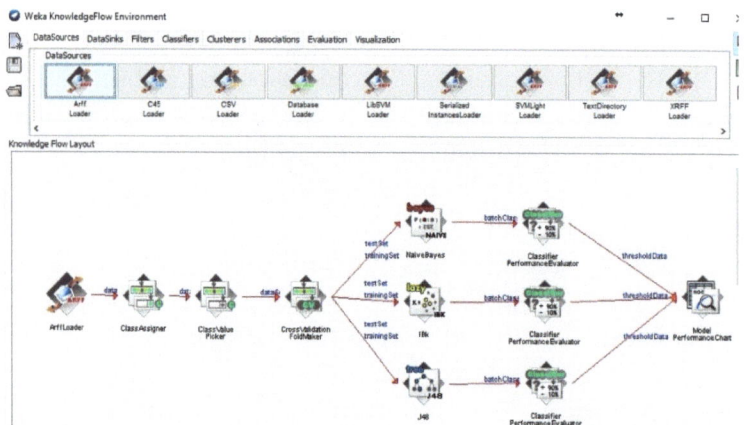

Figure 46 Knowledge flow to evaluate different classifiers by their ROC curves.

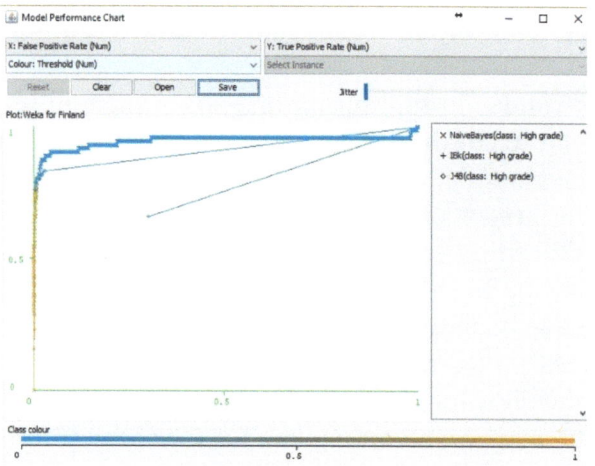

Figure 47. ROC curves for different classifiers.

The best model is the closest to the Y-axis first and then to the X-axis. In our example, the Naïve Bayes gives a slightly better result for the high-grade classification than J48.

Chapter VII
Processing the data

What is CoDaPack?

Over the last several years, a new methodological approach has been developed for the statistical analysis of compositional data, following the approach introduced in the early eighties by John Aitchison (1982). This methodology was difficult to use with standard statistical packages. For this reason, a group of professors from the Girona Compositional Data Group of the University of Gerona has developed The Compositional Data Package, that implements the most elementary of mentioned statistical methods (Comas et al., 2011). This freeware is for users coming from the applied sciences, with no extensive background in using various computer packages. CoDaPack has the following features:
1. Transformations between the real space to the simplex or vice-versa such as the ALR, CLR and ILR transformations.
2. Operations inside the simplex: centering, perturbation, power transformation, amalgamation, sub-composition (closure) or rounded zero replacement.
3. 2-D and 3D graphical outputs like ternary diagrams, ALR plots, CLR plots, biplots, and plots of principal components.
4. Compositional Descriptive Statistics.

Compositional data analysis

To conduct the study of your data using the concepts of compositional data analysis (R. Valls, 2008), do the following steps:
1. Add one column before the last one (Class). Name that column "Rest" and subtract from 100 the sum of all your elements taking into consideration their units. As a check-up, unless your data included major oxides, that number should always be close to 100.

2. Transform the class values to the original gold values.
3. Organize the remaining elements in accordance to the different suites (Oxidation, Base Metals, HFSE, etc.) as shown in Table 9. Leave the REE as just one column

Table 9. Geochemical association of elements.

Oxidation suite	Base Metals	BM: Chalcophile associated indicators	High Field Strength Elements	Rare Earth Elements	Platinum Group Elements	Lithophile Elements	Major oxides
As	Co	Ag	Cr	Ce	Os	Ba	Al
Au	Cu	Bi	Hf	Dy	Pd	Be	Ca
Br	Ni	Cd	Nb	Er	Pt	Cs	Fe
Cl	Pb	Ga	Ta	Eu	Ru	Li	K
Hg	Zn	Ge	Ti	Gd		Mn	Mg
I		In	Y	Ho		Rb	Na
Mo		Sn	Zr	La		Sc	P
Re		Tl		Lu		Sr	S
Sb				Nd			
Se				Pr			
Te				Sm			
Th				Tb			
U				Tm			
V				Yb			
W							

4. Any column with non-numeric data (text) will be used by CoDaPack for groups. Add geology and convert the sample numbers to text.
5. Make sure that your data exclude negative numbers or b.d.l.
6. Open CoDaPack and import your Excel or csv file (Fig. 48).

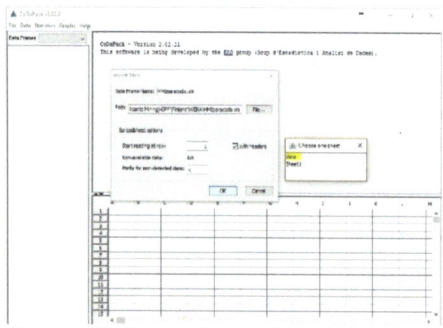

Figure 48. Importing an Excel file into CoDaPack.

7. First, determine the presence of atypical data (Fig. 49).

Exploring Geological Data with Weka, CoDaPack, and iNZight

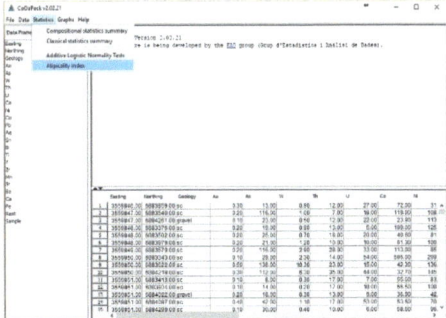

Figure 49. Determining the presence of atypical data.

For that, select all elements (except for the coordinates), adjust the threshold if you want something different than the usual 95% (Fig. 50), and click Accept.

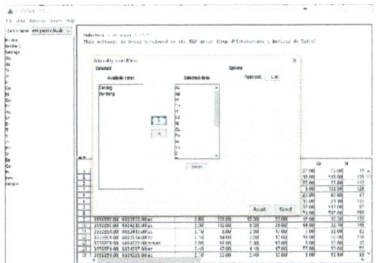

Figure 50. Determining atypical data.

The program will show a list of the atypical samples and it will create an extra column with the atypicality index for each sample. You can later use the Post option of SURFER to post those atypical samples.

8. Run Statistics/Classical statistics summary to get the normal statistics (Fig. 51).

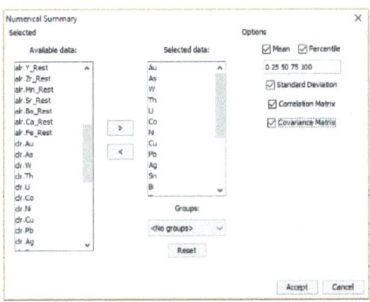

Figure 51. Classical statistics summary.

53

9. Run Statistics/Compositional statistical summary. Focus here on the variation table (Fig. 52). Look for the elements with the largest variances (in red). These elements retain the largest possible variability of the complete data set (As, Ca, Fe, Mn, Sn, Ti, and W).

Figure 52. An example of a variation array produced by CoDaPack.

10. Next, run Graph/CLR biplot (Fig. 53).

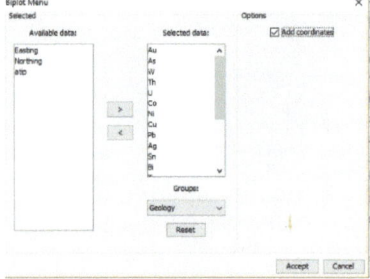

Figure 53. Preparing the data for a biplot CLR analysis.

The CLR biplot is one of the most useful in CoDaPack (Figs. 54-55).

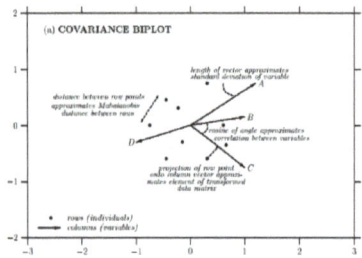

Figure 54. Graphical explanation of the CLR biplot.

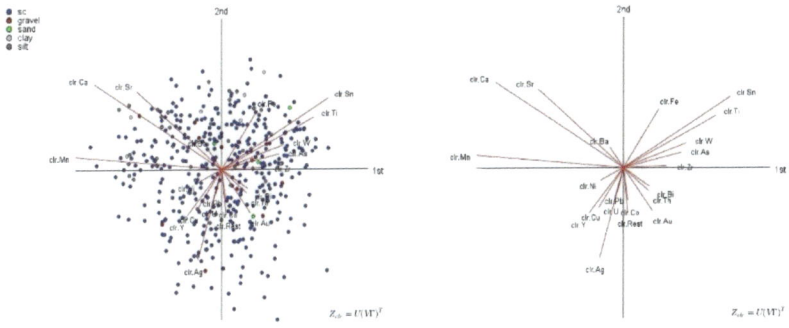

Figure 55. Results of the CLR Biplot.

11. Some interpretation guidelines for the biplot (Aitchison & Greenacre, 2002; Aitchison, Ng, & others, 2005; Vera Pawlowsky-Glahn, Egozcue, & Tolosana Delgado, 2007) follow:
   a. Short rays are inconsequential.
   b. Elements on perpendicular axes have no correlations between them (e.g. Mn-Ni with U-Fe).
   c. Elements close together have high correlation.
   d. Elements close together have redundant information.
   e. Separated elements have higher variability (e.g. Mn-Co).
   f. For long axes, aligned rays (3 or more) are considered to have a linear relationship (e.g. Zr-Bi-Th or Ni-Pb-Co).
   g. For ternary plots, select two axes that are close against and axes that are opposite (e.g. Bi-Th-Sr).

This process also offers a full Principal Component analysis table that you can export to Excel and transpose to obtain the best components (Table 10).

P. Geo. Ricardo A. Valls

Table 10. Principal component analysis of the CLR biplot routine.

Biplot generated:
Data: As Ag W Th U Co Bi Cu Pb Ag Sb Ni Ti Y Zr Mn Sr Ba Ca Fe Rest

Principal Components:

[Table of PCA loadings for components PC1–PC20 across variables clr.Au, clr.As, clr.W, clr.Th, clr.U, clr.Co, clr.Bi, clr.Cu, clr.Pb, clr.Ag, clr.Sb, clr.Ni, clr.Ti, clr.Y, clr.Zr, clr.Mn, clr.Sr, clr.Ba, clr.Ca, clr.Fe, clr.Rest, Cum.Prop.Exp. — values not legibly reproducible from the source image.]

The steps below explain how to work with the data in Excel:
i. Highlight the table on CoDaPack and press Ctrl+C to copy it.
ii. Open Excel and paste the table.
iii. To determine the amount of PCA select the last column and insert a scatter graphic. Select the limit where the curve starts to go horizontal (point 8 in Fig. 56).

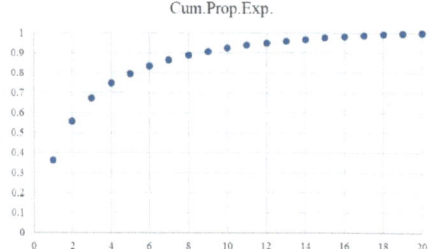

Figure 56 Cumulative explanation of the PCA.

iv. Copy the first 8 PCA from the table[2], click somewhere outside it, and use Paste/Transpose.
v. Select the transposed table and use the conditional formatting to define the top 10% and the lower 10% of the values (Fig. 57).

---

[2] Exclude the last column.

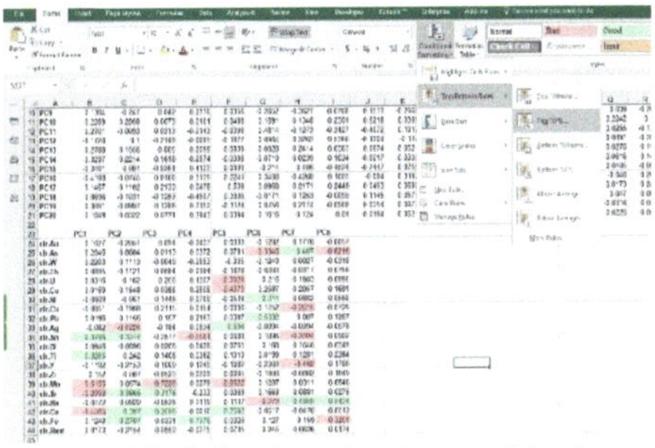

Figure 57. Selecting the top and bottom values for each PCA.

vi. For each PCA, determine the most significant association of elements and look for their geological explanation (Table 11).

Table 11. Obtained PC from the CLR boxplot.

PC	Composition	Explanation
1	(Sn+Ti)/(Mn+Sr+Ca)	Possible alteration
2	(Sn+Sr+Ca+Fe)/Ag	Possible mineralization
3	(Sr+Ca)/Mn	Carbonatization?
4	Fe/Sn	Rhodostannite?
5	(Ag+Ca)/(U+Co+Mn)	Possible mineralization
6	(Ni+Pb)/(As+Ba)	Volcanic indicator?
7	(Ca+W)/(Cu+Sn+Y)	Scheelite? Possible mineralization
8	Ca/(As+Fe)	Scorodite? Hydrothermal alteration

Biplot to select informative elements

You now know from Fig. 55 that longer axes indicate larger variability, so from Figure 58 you probably selected Mn, Ca, Sr, Sn, and Ag in the first order and in second order elements like Fe, Ti, W, Au, and Y.

Remember that vectors that are close together, e.g. Bi and Th, Pb-Co-U, or Cu and Y, have strong correlations, therefore they provide the same amount of information (they are redundant).

For ternary analysis, look for two elements that are close together with one that is opposite to them. For example, Ca, Sr, and Bi. Select Ternary Principal Components from the Graphs Menu and then select those three or four elements and a Group if desired, and then click Accept. The ternary or tetragon diagram will be shown in a separate window (Fig. 59) and the results of the PCA will be in the information window in CoDaPack.

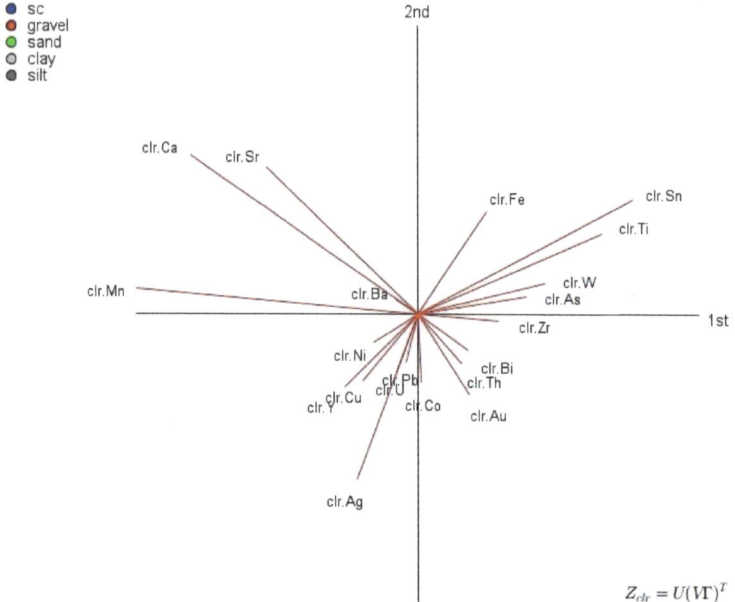

Figure 58. Using a CLR biplot to determine the informational value of the different elements.

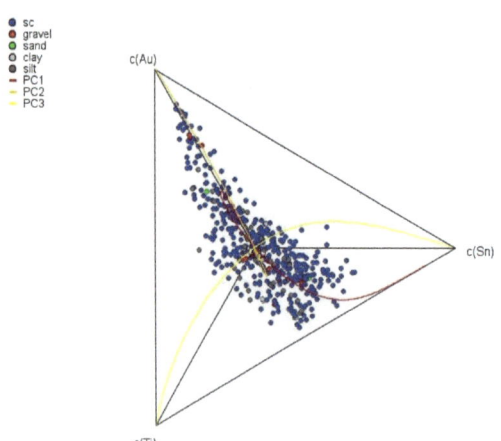

Figure 59. Tetragon diagram for Au, Ni, Sn, and Ti.

Balance dendrogram

The dendrogram is a graphic representation of the cluster analysis (Fig. 60).

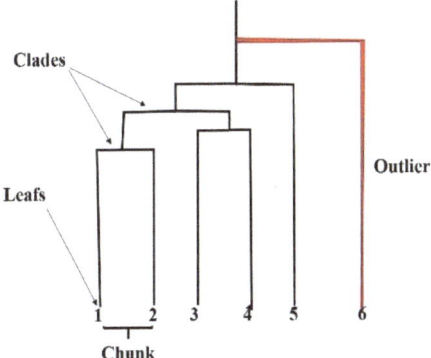

Figure 60. Parts of a dendrogram.

The interpretation of a dendrogram is straightforward. Starting from the top of the dendrogram down, the clades define groups of associated parameters. The length of the leaf indicates how similar are the chunks within a clad and need to be interpreted from bottom to top.

For example, in Fig. 60 there are five clades, one of them is an outlier (6), one is composed of just one parameter (5) and the third one is more complex, composed of two subgroups of clades each with two chunks. Parameters 1 and 2 in the first chunk are more like each other than parameters 3 and 4 in the next chunk.

To obtain a dendrogram using CodaPack:
1. Select "Balance dendrogram" under Graphics.
2. Select all elements, including the column "Rest".
3. Select the partition table in accordance with the enzyme leach association of elements (Table 9) or any other associative combination you desire to test[3].

---

[3] I explain how to do the manual partition in "Definition of a partition" under ILR transformations on page 67.

I have found that if you separate the data in classes, for example barren and interesting, the dendrogram shows relationships that are harder to see than if you treat the entire data set (Figs. 61-62).

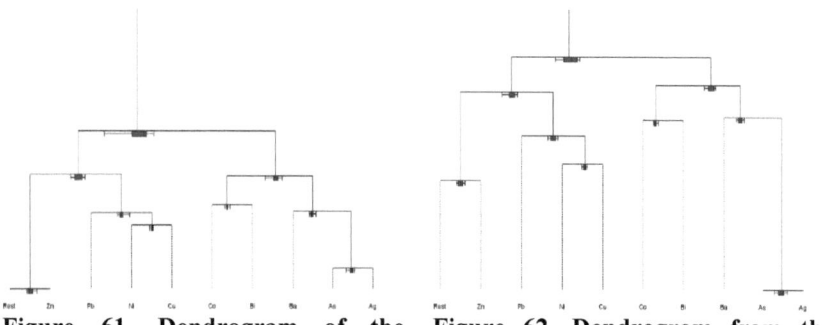

Figure 61. Dendrogram of the barren data from an MMI study.

Figure 62. Dendrogram from the interesting data from the same MMI study.

While the determined clads are similar[4], the lengths of the leaves are different. The chunk As-Ag is shorter for the "interesting" data than for the barren ones and the chunk Rest-Zn is shorter for the barren data than for the interesting data.

You can also analyze the transposed values of the original data with a dendrogram. Use the transpose option under paste on Excel to convert the columns to rows. Then repeat the dendrogram using CoDaPack. Fig. 63 shows the results for the interesting data set.

---

[4] The reason why the clads are similar is that they respond to the same partition.

Exploring Geological Data with Weka, CoDaPack, and iNZight

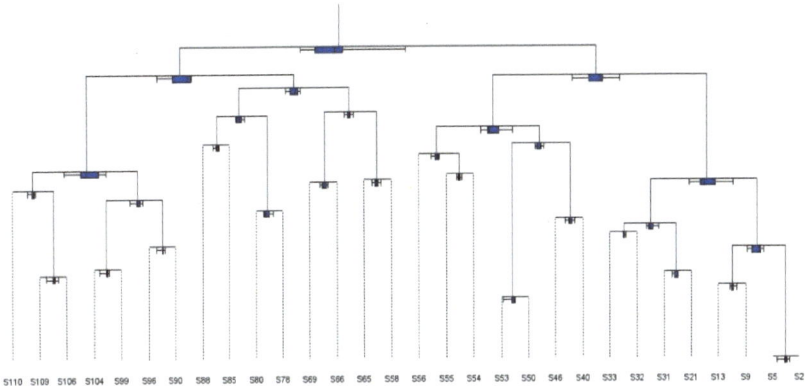

Figure 63. Inverse dendrogram for the interesting data of the MMI survey.

The idea here is to represent on a map the most similar chunks, e.g. S5 and S2, S53 and S50, S13 and S9, etc. and to interpret the geological implication of such chunks or clads.

A final word on dendrograms. Many statistical programs will create dendrograms. For example, MYSTAT creates nice graphics and provides numeric data on the length of the leaves (Fig. 64). I recommend using CoDaPack instead in order to eliminate the closure problem of the data.

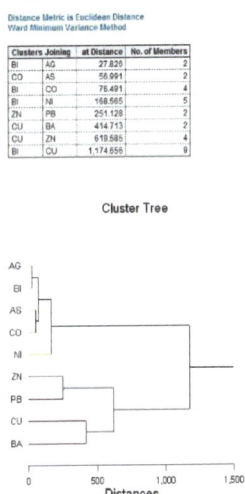

Figure 64. Dendrogram obtained with MYSTAT free software.

Transforming your original data

CoDaPack transforms the original data to ALR, CLR, and ILR data (Thió-Henestrosa & Comas-Cufi, 2011). Any of these algorithms will open the close data and allow you to use normal statistical procedures (Fig. 65).

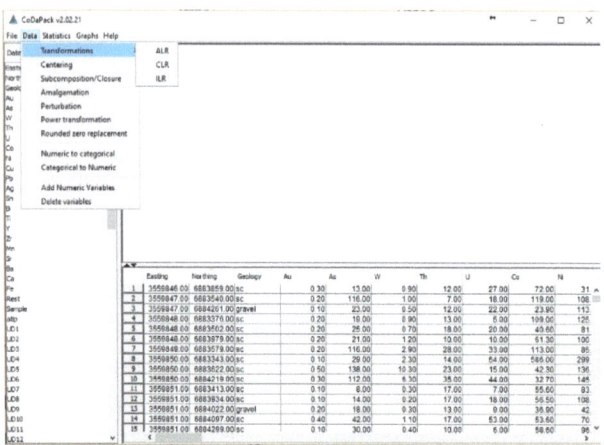

Figure 65. Transforming your RAW data.

ALR

With this option (Fig 66) the data is transformed with the additive log-ratio transformation (ALR) (Aitchison, 1982) from the simplex (raw data) to real space (ALR coordinates)

$$y = alr(x) = \left[\ln\frac{x_1}{x_D}, \ln\frac{x_2}{x_D}, \ldots, \ln\frac{x_{D-1}}{x_D}\right],$$

where y ∈ R^{D-1}, the (D − 1)-dimensional real space, or with its inverse, the generalized additive logistic transformation (agl),

$$x = agl(y) = \left[\frac{\exp(y_1)}{1+\sum_{i=1}^{D-1}\exp y_i}, \ldots, \frac{\exp(y_{D-1})}{1+\sum_{i=1}^{D-1}\exp y_i}, \frac{1}{1+\sum_{i=1}^{D-1}\exp y_i}\right],$$

from the real space (ALR coordinates) to the simplex (raw data). In the ALR transformation, the divisor is taken to be the last part according to the sequence selected by the user. The interface allows the user to reorder the variable once they have been selected by dragging its name in the selected data list. The ALR coordinates depend on the divisor and they conform an oblique basis (Egozcue & Pawlowsky-Glahn, 2005).

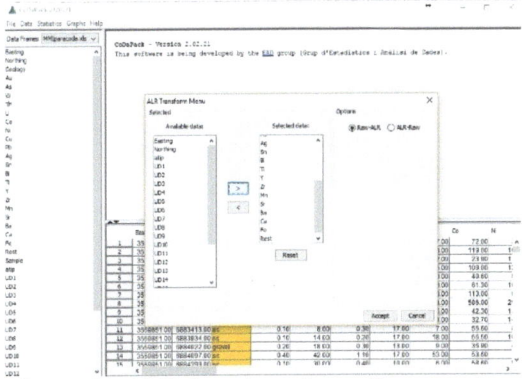

Figure 66. ALR data transformation.

Remember to add the column "Rest" to the ALR selected data window.

CLR

With this feature (Fig. 67) the data is transformed from the simplex (raw data) to real space (CLR-coefficients) according to the centered log ratio transformation (CLR)

$$y = \text{clr}(\mathbf{x}) = \left[\ln \frac{\mathbf{x}}{g_D(\mathbf{x})}\right] = \left[\ln \frac{x_1}{g_D(\mathbf{x})}, \ln \frac{x_2}{g_D(\mathbf{x})}, \ldots, \ln \frac{x_D}{g_D(\mathbf{x})}\right],$$

where $y \in R^{D-1}$ and $g_D(x)$ is the geometric mean of the parts involved, i.e.

$$g_D(\mathbf{x}) = \left(\prod_{i=1}^{D} x_i\right)^{1/D} = \exp\left(\frac{1}{D}\sum_{i=1}^{D} \ln x_i\right),$$

or with the inverse transformation (CLR−1), from real space (CLR coefficients) to the simplex (raw data).

$$\mathbf{x} = \text{clr}^{-1}(\mathbf{y}) = \left[\frac{\exp(y_1)}{\sum_{i=1}^{D}\exp y_i}, \frac{\exp(y_2)}{\sum_{i=1}^{D}\exp y_i}, \ldots, \frac{\exp(y_D)}{1+\sum_{i=1}^{D}\exp y_i}\right].$$

The CLR coordinates represent a generating system, not a basis, and therefore CLR coordinates sum up to zero (Egozcue & Pawlowsky-Glahn, 2005). Therefore, covariances and correlations between CLR-parts have the same drawbacks as covariances and correlations between compositional parts: their sub compositions are incoherent.

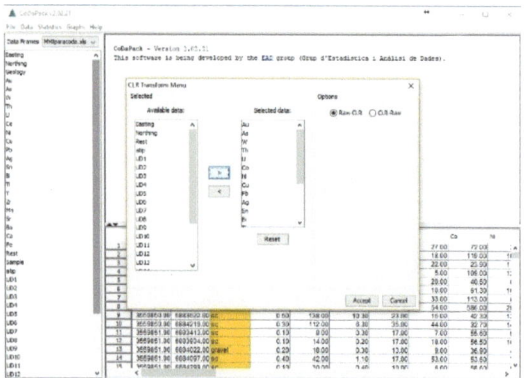

Figure 67. CLR data transformation.

ILR

With this feature (Fig. 68) the data is transformed from the simplex (raw data) to real space (ILR coordinates) with the isometric log ratio transformation (ILR), or from real space to the simplex applying the inverse isometric log ratio transformation (ILR−1), both defined by a sequential binary partition (Egozcue, Pawlowsky-Glahn, Mateu-Figueras, & Barcelo-Vidal, 2003)

The ILR transformation is represented by $y = ILR(x) = (y1, y2, ..., y_{D-1}) \in R^{D-1}$, where $yi = \sum_{j=1}^{D} \psi ij \ln xj$, $i = 1, 2, ..., D-1$, and

$$\psi_{ij} = \begin{cases} \sqrt{\frac{s_i}{r_i(s_i + r_i)}}, & \text{if at step } i \text{ the part } j \text{ is coded in the SBP as } +1; \\ -\sqrt{\frac{r_i}{s_i(s_i + r_i)}}, & \text{if at step } i \text{ the part } j \text{ is coded in the SBP as } -1; \\ 0, & \text{if at step } i \text{ the part } j \text{ is coded in the SBP as } 0; \end{cases}$$

with r_i the number of parts coded at step i in the SBP as +1, and s_i the number of parts coded at step i in the SBP as −1.

The ILR^{-1} transformation consists on: $x = ILR-1(y) = (x1, x2, ..., xD) \in SD$, where $[x1, x2, ..., x_D] = C \exp[z1, z2, ..., z_D]$, $zj = \sum_{j=1}^{D-1} \psi ijyi$, C stands for the closure operation.

Exploring Geological Data with Weka, CoDaPack, and iNZight

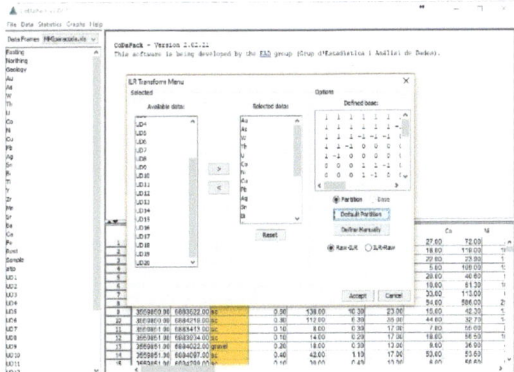

Figure 68. ILR data transformation.

Definition of a partition

A partition is a hierarchical grouping of parts of the original compositional vector, starting with the whole composition as a group and ending with each part in a single group (Egozcue & Pawlowsky-Glahn, 2005). First, the compositional vector is divided into two non-overlapping groups of parts. In a similar way, each of these two groups is divided again, and so on until each group contain only a single part. If D is the number of parts of the original composition, the number of steps of the partition is D−1. CoDapack includes two different ways to define a partition:

1. Default partition. The default partition is defined by the Haar basis[5]. It consists in separating, at each step, the parts approximately in the middle.
2. Defined by the user. By activating this option, a new button appears and clicking on it will show a new window. This window (Fig. 69) has a grid where rows represent parts and columns the steps of the partition.

To define the partition, every time you mark one part with a single click, a + sign appears in the grid at the cell corresponding to this part in the current step. At each step of partition, a + sign

---

[5] The Haar wavelet is a sequence of rescaled "square-shaped" functions which together form a wavelet family or basis.

means that the part is assigned to the first group, a − sign to the second, and it remains blank if this part is out of the group that is divided in this order. To remove a + sign from the current stepm, it is necessary to mark the cell of the current step of the partitioning grid that contains this + sign with a single click. To finish a step, press the Next Step button.

At each step, it is only possible to divide one group. This group is marked with a green color on the partitioning grid. To facilitate this task, when the Next Step button is pressed, all the information (labels and partition) is reordered in such a way that the next parts to divide appear in a sequence. To eliminate some steps of the partition, press the Previous Step button as many times as required.

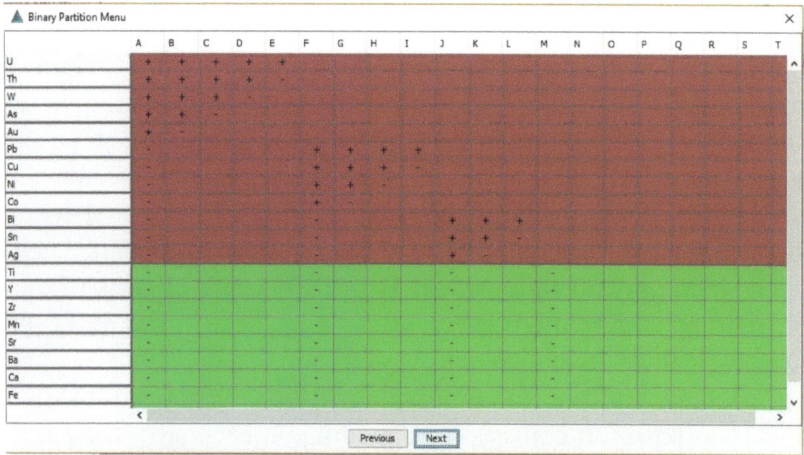

Figure 69. Defining a manual partition for ILR transformation.

You could use the default option, but the manual option gives a unique opportunity to include geological information in the interpretation of the data.

"Normal" statistics

Now that you have converted your raw data with these algorithms, apply your usual statistical methods like covariance, correlation, regression analysis, etc. These are part of the Data Analysis function of Excel.

For more advanced methods for example cluster analysis, principal component analysis, etc., use a more specialized software.

I currently use MYSTAT that is a free, streamlined, student-oriented variation of the SYSTAT 13 flagship product, featuring statistical routines that are covered in undergraduate-level statistics, science, and social science courses. You can download the software from- http://ow.ly/VJAp30e8I35.

I also use WINSTAT, a low-priced ($99, free demo) and easy-to-use solution for a range of statistical analyses on Excel data. Some functions include:

- Regression analysis: linear regression, multiple regression, polynomial regression.
- Correlation: Pearson correlation, Spearman correlation, partial correlation.
- Analysis of variance (ANOVA).
- Repeated measures ANOVA.
- Survival analysis (Kaplan-Meier).
- Cox regression.
- Discriminant analysis.
- Cluster analysis.
- Factor analysis.
- t-Tests: dependent t-test, independent t-test.
- Mann-Whitney U-test.
- Wilcoxon test.
- Kruskal-Wallis H-test.
- Box-plot, scatter plot, histogram, probit chart.
- Crosstabs.
- Fisher's exact test
- Kolmogorov-Smirnov, Shapiro-Wilk, Chi-square test.
- McNemar.

- Tests of normal distribution.
- Tests of randomness.
- Outlier tests.
- Pareto diagram, Quality control charts, Process capability.

You can download a 30-days fully functional demo from-http://ow.ly/YZHn30e8ItX.

Keep a record of all the procedures that you do to your original raw data, so that anyone can verify your results. Many people use R programming for this, but you can achieve the same result using Excel (Fig. 70).

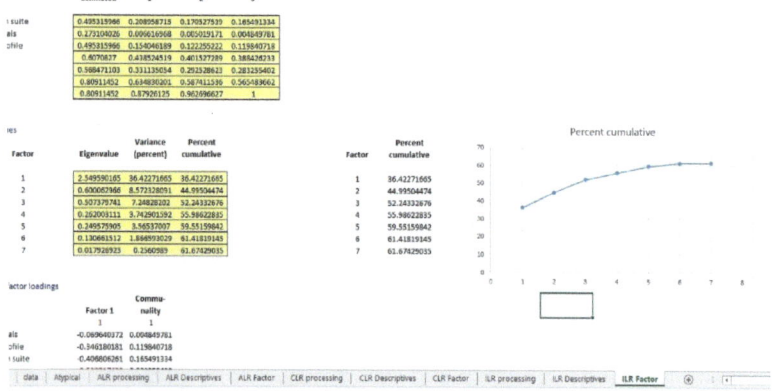

Figure 70. The different tabs in Excel record all the procedures.

Scatterplots

Use ALR, CLR, or ILR transformed data with Excel to construct scatterplots (Fig. 71).

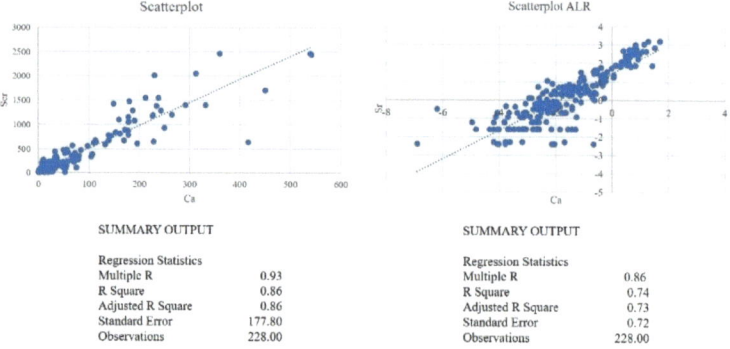

Figure 71. Comparing scatterplots.

Ternary principal components

Once you have determined the appropriate ternary elements using the CLR biplot (e.g. Ca-Sr-Bi), use CoDaPack to obtain their ternary principal component analysis (Fig. 72).

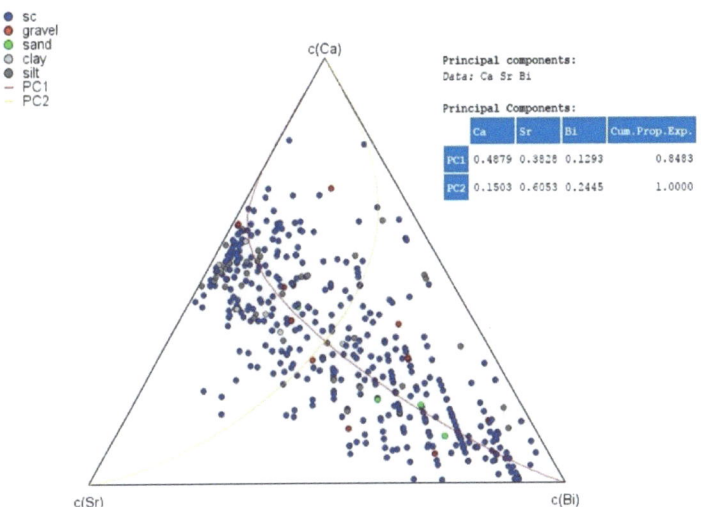

Figure 72. Ternary principal component analysis.

As you can see from Fig. 72, the first component explains 84% of the total variability of the data.

Chapter VIII
Presenting the data

I will show some ways of presenting the data that will allow you to extract more information from the data. Remember to use the transformed data (ALR, CLR, and ILR) to avoid the closure problems.

INZight

This software was first designed for New Zealand high schools, allowing students to explore data and understand some statistical ideas. Now it extends to multi variable graphics, time series, and generalized linear modeling. The software is free and you can even modify it if you are so inclined. However, note that iNZight is unwarranted. You can download the program online from http://ow.ly/hnwl30evYRt and you can access the online help for this program at http://ow.ly/KlnO30evYTj.

To explain the capabilities of this software, I will use the data set "MMI transformed". You can download it from the web at http://ow.ly/l0Lr30eoSE1.

While iNZight can open many types of files, I found that it works better importing csv files. Once you import the data, save it as an iNZight file (.xrd).

I will show ways to represent results from one, two, and three or more elements. You can find a detailed user guide for iNZight under Help (Fig. 73).

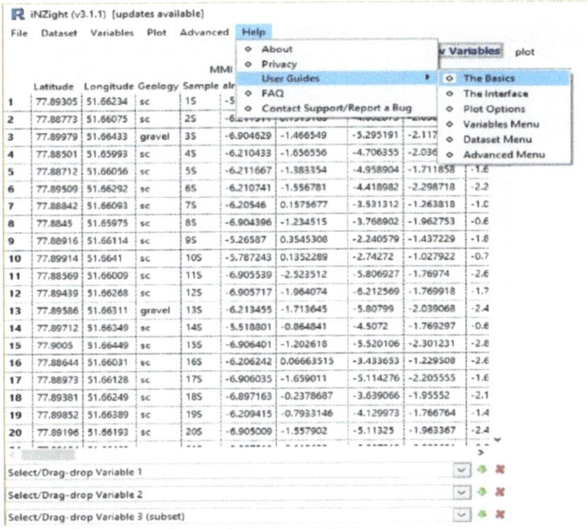

Exploring Geological Data with Weka, CoDaPack, and iNZight

Figure 73. User guides for Inzight.

Mono element representation

You can drag and drop an element from the labels to the spot for Variable 1, or select that element from the drop-down menu. INZight will select the proper type of graphic. For example, if you select the parameter "geology", Inzight will create a bar chart (Fig 74), but if you select a numeral parameter, e.g. ALR.Au, the software will create a dot graph with a Wilkinson box below (Fig. 75).

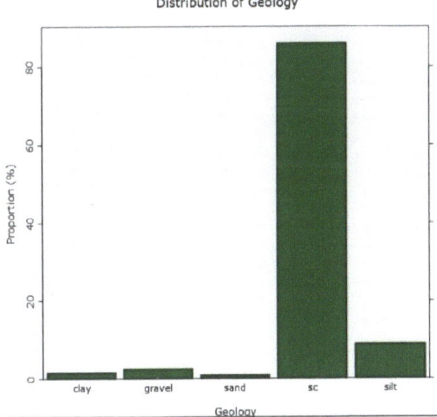

Figure 74. Bar chart (histogram) for the distribution of the geology.

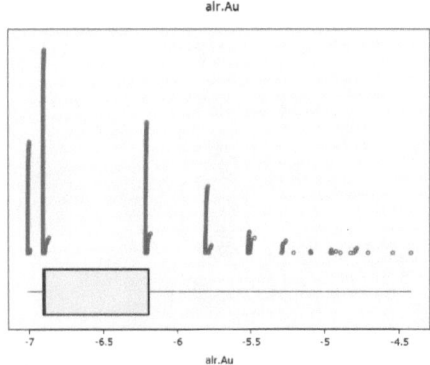

Figure 75. Dot bar for the ALR.Au.

If you go back to the parameter "geology" and select "Get summary", you will get the numeric parameters of the distribution of the parameter "geology" (Fig. 76).

```
R Summary
===============================================================
                          iNZight Summary
---------------------------------------------------------------
Primary variable of interest: Geology (categorical)

Total number of observations: 436
===============================================================

Summary of the distribution of Geology:
---------------------------------------
              clay    gravel     sand       sc      silt    Total
Count            7        11        4      375        39      436
Percent     1.6055%   2.5229%  0.9174%  86.0092%  8.9450%    100%

===============================================================
```

Figure 76. Inzight summary for the parameter geology.

If you select "Get inference" you will get more numeric information from the bar chart, including nonparametric bootstrap estimations (Fig. 77).

Exploring Geological Data with Weka, CoDaPack, and iNZight

```
R Inference Information
================================================================
                     iNZight Inference using the Nonparametric Bootstrap
----------------------------------------------------------------
  Primary variable of interest: Geology (categorical)

  Total number of observations: 436
================================================================

Inference of the distribution of Geology:
----------------------------------------

Estimated Proportion with 95% Percentile Bootstrap Confidence Interval

         Lower    Estimate    Upper
  clay   0.00459  0.01597    0.0298
 gravel  0.01147  0.02516    0.0413
  sand   0.00229  0.00918    0.0183
   sc    0.82798  0.86035    0.8922
  silt   0.06422  0.08935    0.1193

### Differences in proportions of Geology
   (col group - row group)

Estimates
           clay      gravel     sand       sc
 gravel   0.009119
  sand   -0.007020   0.83489
   sc   -0.016139   0.85103   0.06434
  silt   0.844009   0.07346   0.08048   -0.7705

95% Percentile Bootstrap Confidence Intervals

           clay      gravel     sand       sc
 gravel  -0.034400   0.81302   0.03440
          0.000000   0.88760   0.09633
  sand   -0.022940   0.79358
          0.009174   0.87390
   sc    0.802750   0.04587   0.05505  -0.8234
          0.880734   0.10440   0.11009  -0.7110
  silt  -0.010380
          0.029817
================================================================
```

Figure 77. Inzight inference using the nonparametric bootstrap for the geology parameter.

If you go back to the ALR.Au and select "Get Summary", you will get a series of standard statistics for the analyzed parameter (Fig. 78). You can also complete the One Sample t-test of the data if you select "Get Inference".

```
R Summary
================================================================
                         iNZight Summary
----------------------------------------------------------------
  Primary variable of interest: alr.Au (numeric)

  Total number of observations: 436
================================================================

Summary of alr.Au:
------------------

   Min     25%    Median    75%     Max     Mean     SD    Sample Size
  -7.012  -6.906  -6.894  -6.191   -4.42  -6.425   0.6126      436
================================================================
```

Figure 78. Basic statistics for the ALR.Au.

If you click on the variable 1 and use your arrows up and down, you can quickly browse through all the graphics from all your parameters. That gives you a great first view of your data.

You can do amalgamation, rank, convert category to numeric and vice versa, and many other procedures within the table. You will find them under "Variables".

You can get the statistics for all your parameters by clicking on the tab Advance/Quick explorer/All 1 variable summaries. You can also get all the graphics following Advance/Quick explorer/All 1 variable plots.

You can see the spatial distribution of your data under Advanced/Maps (Fig. 79).

Figure 79. An example of a map constructed with Inzight.

Bi element representations

When you select two elements on iNZight, the program shows you a scatter plot type of diagram. By clicking on the "Add to plot" icon (first of the three at the bottom right part of the window) you can customize the graphic by:
1. Coloring in dependence of a variable.
2. Editing the size and type of each point.
3. Adding trend lines and curves.
4. Identifying statistical outliers (Fig. 80).

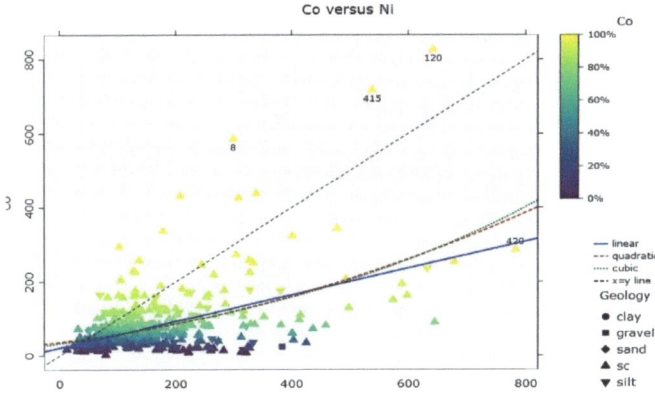

Figure 80. Scatter plot of the data. Note the statistical outliers 8, 415, 120, and 420.

By pressing "Get Summary" and/or "Get Inference" you can get all the numeric data of the different curves.

You can get more detailed scatter plots with the tab Advanced/Quick Explore/Explore 2 variables plots or Pairs (Fig. 81).

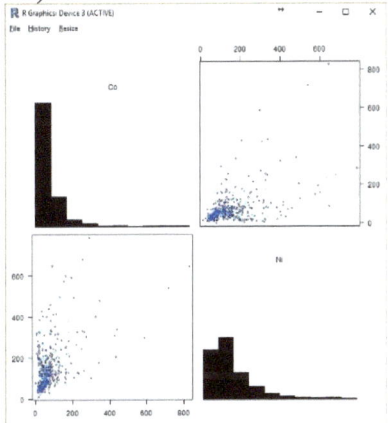

Figure 81. The combination of histograms and scatter plots.

Another option under the tab Advanced is the Model Fitting (Fig. 82), where you can find the model to estimate one element (Y) by any combination of other elements, apart from a complete set of different graphical representations (Fig. 83), normality tests (Fig. 84), etc.

Figure 82. Model fit options in iNZight.

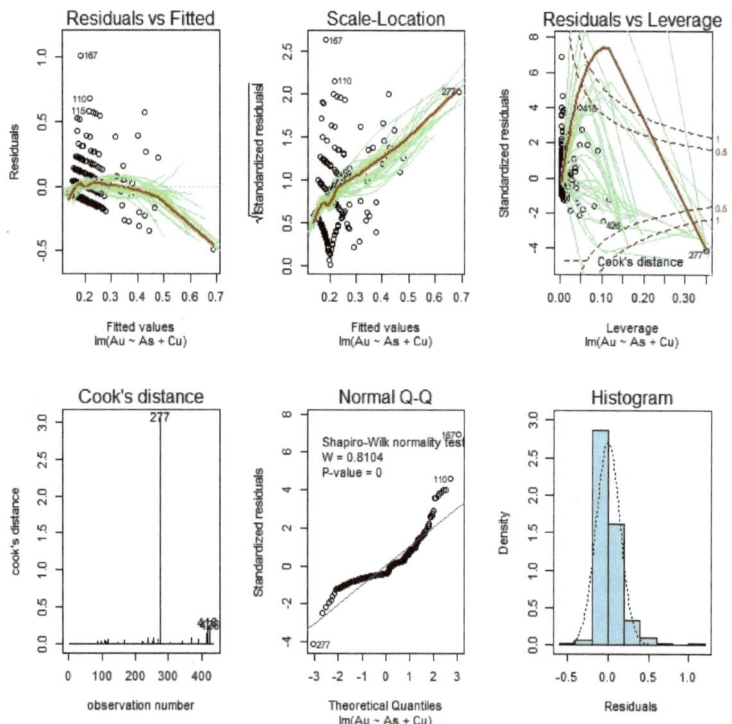

Figure 83. Summary of the different plots for the Fit Model.

Exploring Geological Data with Weka, CoDaPack, and iNZight

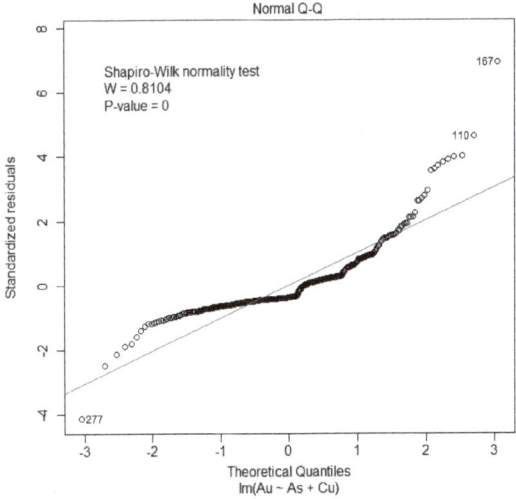

Figure 84. The normal Q-Q curve for the estimated model.

Multielement representations

Three elements

INZight can process up to four variables at a time. It can explore the relationship between two elements or parameters controlled by a third one. Fig. 85 shows the relationship (scatterplots) between the oxidation suite and the base metal suites grouped by the geology.

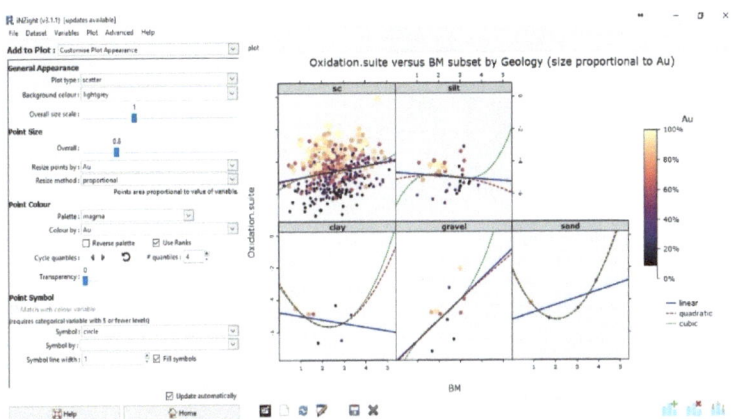

Figure 85. An example of a three-parameter modeling.

If you select the "_MULTI" option on the Home page, you can see the individual maps in a dynamic sequence.

All the previously explained options will also work for these graphics.

You can customize a 3D dynamic graphic under Advanced/3D Plot (Fig. 86) and then you can click on Plot in #D to obtain a dynamic graphic (Fig 87).

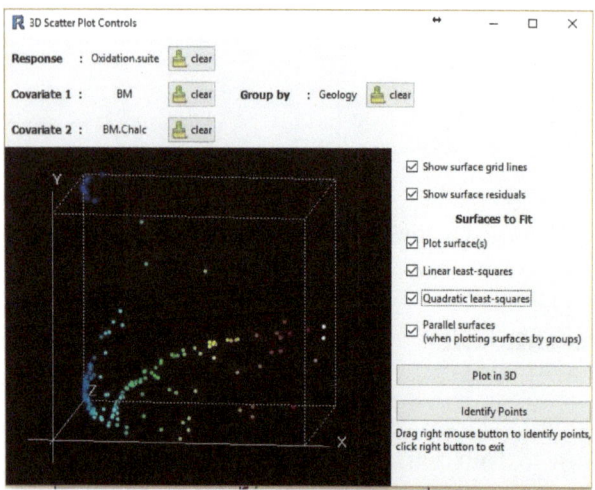

Figure 86. Formatting the 3D representation of the data.

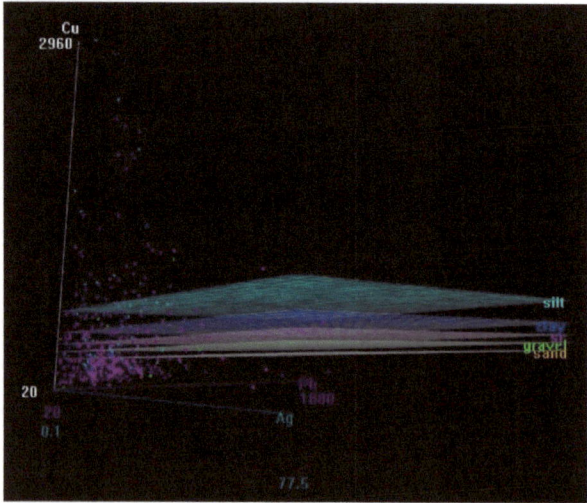

Figure 87. An example of the 3D dynamic representation of the data.

Exploring Geological Data with Weka, CoDaPack, and iNZight

Four elements

INZight gives you the possibility of observing 4 variables at the same time. In the example shown in Fig. 88, you can analyze the scatterplots between Au and As by subsets of values of W, grouped by the geology.

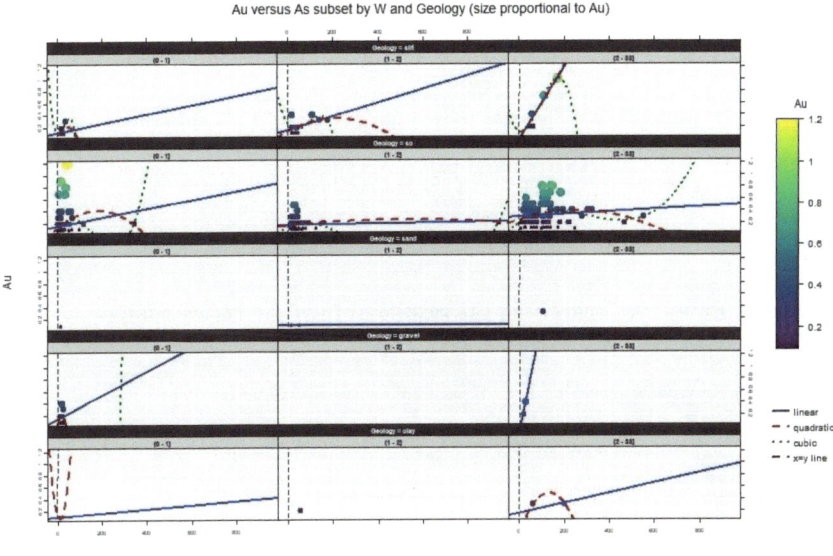

Figure 88. Graphic representation of 4 parameters with Inzight.

All the same capabilities of adding lines and getting summaries and inference data as explained before work here too, as well as the possibility of obtaining dynamic representations by selecting the _MULTI or the _ALL options from the homepage.

Time series

Although they are infrequently used in geochemistry, iNZight can represent time series (Figs. 89-90) and even predict future results (Fig. 91).

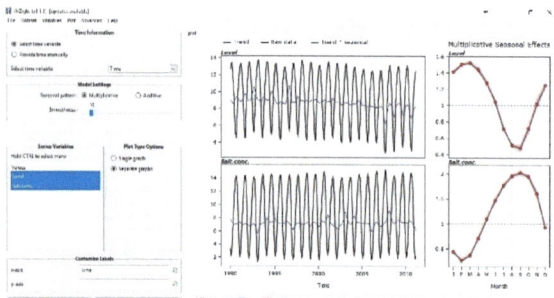

Figure 89. Panel for formatting the results of the time-series analysis.

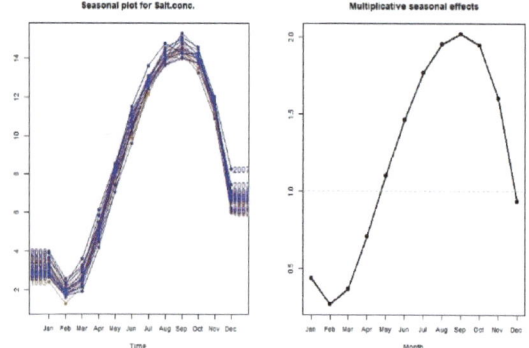

Figure 90. An example of the seasonal plot for an element in iNZight.

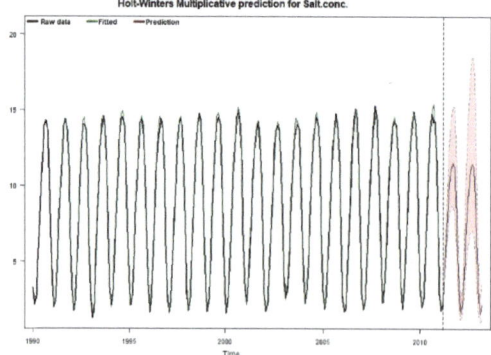

Figure 91. Using iNZight to predict the behavior of the data in the future.

Exploring Geological Data with Weka, CoDaPack, and iNZight

Other free programs to consider

I would like to introduce you to a couple of free online sources to graphically represent data in original ways. These sources are Plotly (Plot.ly) and DataHero (app.datahero.com).

They allow users to upload their data from any browser and to process and obtain useful graphics that can be downloaded to your machine (Figs. 92-93).

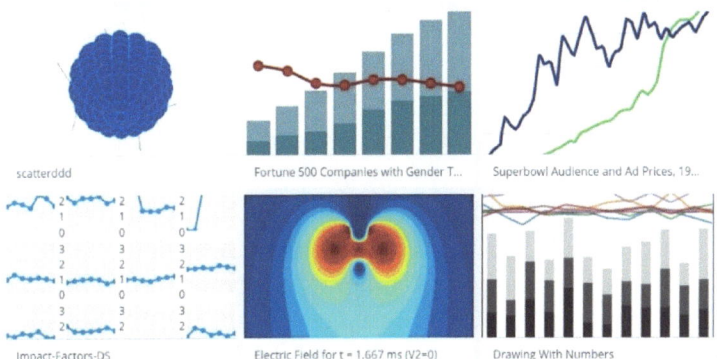

Figure 92. Examples of graphics from Plotly.

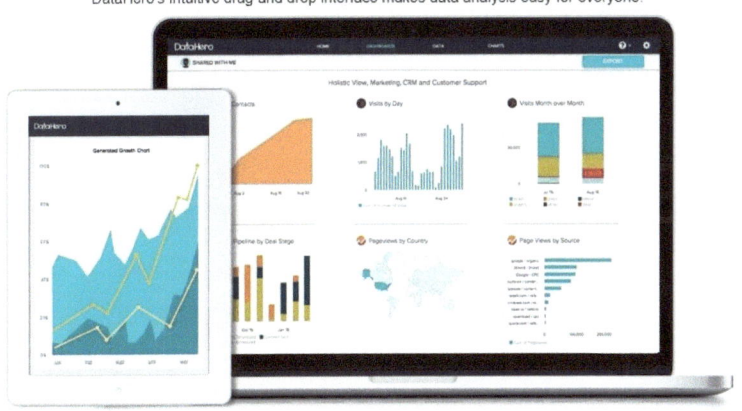

Figure 93. Examples of graphics from DataHero.

Chapter IX
Conclusions and recommendations

Most of the geological and geochemical data is closed and to extract valid information from the data it is mandatory to apply compositional data analysis. CoDaPack is an Excellent program to process closed data and I strongly recommend its use.

Weka is a useful and user-friendly software that processes "big data", stream data, and smaller datasets. In the preceding pages I presented a step-by-step illustrated guide to explain how to use both Weka and CoDaPack when dealing with geochemical data. These methods can and should be applied to non-geochemical data sets as well.

I discussed the most common features of iNZight and how to use it to process and present data in ways that facilitate their interpretation. I also mentioned other free online software to provide you with a wider variety of choices. You can explore your data using adds-on like "Data Tool", "Statistics", and other from Google Sheet (https://docs.google.com) that also generate batches of graphics and are useful for the first look at your data.

You can find the datasets used in this book here- http://ow.ly/l0Lr30eoSE1.

References

Aitchison, J. (1982). The statistical analysis of compositional data. *Journal of the Royal Statistical Society. Series B (Methodological)*, 139–177. JSTOR.

Aitchison, J. & Greenacre, M. (2002). Biplots of compositional data. *Journal of the Royal Statistical Society: Series C (Applied Statistics)*, *51*(4), 375–392. Wiley Online Library.

Aitchison, J., Ng, K. & others. (2005). Conditional compositional biplots: theory and application. Universitat de Girona. Departament d'Informàtica i Matemàtica Aplicada.

Bacon Shone, J. & others. (2003). Modelling structural zeros in compositional data. Universitat de Girona. Departament d'Informàtica i Matemàtica Aplicada.

Comas, M., Comas-Cufi, M. (Universidad de G., Thió-Henestrosa, S. (Universidad de G., Egozcue, J., Tolosana-Delgado, R. & Ortego, M. (2011). CoDaPack 2.0: a stand-alone, multi-platform compositional software. *Options* (pp. 1–10).

Egozcue, J. J. & Pawlowsky-Glahn, V. (2005). Groups of parts and their balances in compositional data analysis. *Mathematical Geology*, *37*(7), 795–828. Springer.

Egozcue, J. J., Pawlowsky-Glahn, V., Mateu-Figueras, G. & Barcelo-Vidal, C. (2003). Isometric logratio transformations for compositional data analysis. *Mathematical Geology*, *35*(3), 279–300. Springer.

Frank, E., Hall, M. & Witten, I. (2016). The Weka Workbench. *Online Appendix for "Data Mining: Practical Machine Learning Tools and Techniques"*, 4th edn. Morgan Kaufman, Burlington.

Kashdan, A. B., Guskov, O. I. & Chimanskii, A. A. (1979). *Mathematical modelling in geology and exploration work* (p. 168). Nedra.

Martín Fernández, J. A., Barceló i Vidal, C., Pawlowsky-Glahn, V., Kovács, L., Kovács, G. P. & others. (2003). Major-elements trends in cenozoic volcanites of Hungary. Retrieved from http://dugi-doc.udg.edu:8080/handle/10256/679

Martín-Fernández, J. A., Palarea-Albaladejo, J. & Olea, R. A. (2011). Dealing with zeros. *Compositional data analysis: Theory and applications*, 47–62. Retrieved from https://books.google.ca/books?hl=en&lr=&id=s9L5TrM1gYEC&oi=fnd&pg=PT65&dq=dealing+with+zeros+and+missing+data&ots=BPJHy7adjM&sig=qRQGJALtTGohusSC05EeBH-qSwg

Pawlowsky-Glahn, V. & Egozcue, J. (2006). Compositional data analysis in the geosciences: from theory to practice. *Geological Society, chapter Compositional data and their analysis: An introduction*, 1–10. Retrieved from http://sp.lyellcollection.org/content/264/1

Pawlowsky-Glahn, V., Egozcue, J. J. & Tolosana Delgado, R. (2007). Lecture notes on compositional data analysis. Universitat de Girona.

Pearson, K. (1896). Mathematical contributions to the theory of evolution. On a form of spurious correlation which may arise when indices are used in the measurement of organs. *Proceedings of the Royal Society of London*, *60*(359-367), 489–498. The Royal Society.

Petrelli, M., Poli, G., Perugini, D. & Peccerillo, A. (2005). PetroGraph: A new software to visualize, model, and present geochemical data in igneous petrology. *Geochemistry, Geophysics, Geosystems*, *6*(7). Wiley Online Library.

Thió-Henestrosa, S. & Comas-Cufí, M. (2011). CoDaPack v2 USER's GUIDE.

Trusova, I. F. & Chernov, V. I. (1982). Petrography of magmatic and metamorphic rocks, Moscow, Nedra, 272 pp.

Valls, R. A. (1987). Tercera Versión de las Fichas Codificadas.

Valls, R. A. (2013). Análisis comparativo entre la vía húmeda y seca de preparación de las muestras de sedimentos, Valls Geoconsultant.

Valls, R. A. (2016). Quality Assurance and Quality Control for the Field Work, Valls Geoconsultant, Toronto, 58 pp.

About the author

As a professional geologist with thirtyfour years in the mining industry, I have extensive geological, geochemical, and mining experience, managerial skills and a solid background in research techniques, and training of technical personnel. I am fluent in English, French, Spanish, and Russian. I have been involved in various projects worldwide (Canada, Africa, Russia, Indonesia, the Caribbean, Central and South America). Projects included regional reconnaissance, local mapping, diamond drilling and RC-drilling programs, open pit and underground mapping and sampling, geochemical sampling and interpretation, and several exploration techniques pertaining to the search for diamonds, PGM, gold, nickel, silver, base metals, industrial minerals, oil & gas, and magmatic, hydrothermal, porphyritic, VMS and SEDEX ore deposits. Special strengths are related to the acquisition of new properties, geochemical and geological studies, management and organization, geomathematical analysis and modeling, compositional data analysis, structural studies, database design, QA&QC studies, exploration studies and writing technical reports. P.Geo. registered in the province of Ontario.

www.ingramcontent.com/pod-product-compliance
Lightning Source LLC
Chambersburg PA
CBHW041101180526
45172CB00001B/53